THE HEATED DEBATE

Greenhouse Predictions Versus Climate Reality

By
Robert C. Balling, Jr.

Introduction by
Aaron Wildavsky

PACIFIC RESEARCH INSTITUTE FOR PUBLIC POLICY
San Francisco, California

ISBN 0-936488-47-6 (cloth)
ISBN 0-936488-48-4 (paper)

Printed in the United States of America.

Pacific Research Institute for Public Policy
177 Post Street
San Francisco, CA 94108
(415) 989-0833

Library of Congress Cataloging-in-Publication Data

Balling, Robert C.
    The heated debate : greenhouse predictions versus climate reality / by Robert C.
    Balling, Jr.
        p.   cm.
    Includes bibliographical references and index.
    ISBN 0-936488-47-6 (cloth) : $21.95
    ISBN 0-936488-48-4 (paper) : $14.95
    1. Climatic changes. 2. Greenhouse gases—Environmental aspects.
    I. Title.
    QC981.8.C5B35   1992
    551.6—dc20   91-29747
                                                                        CIP

CONTENTS

LIST OF FIGURES

LIST OF TABLES

GLOBAL WARMING AS A MEANS OF ACHIEVING AN EGALITARIAN SOCIETY: AN INTRODUCTION

Global warming is the mother of environmental scares. In the scope of its consequences for life on planet Earth and the immense size of its remedies, global warming dwarfs all the environmental and safety scares of our time put together. Warming (and warming alone), through its primary antidote of withdrawing carbon from production and consumption, is capable of realizing the environmentalist's dream of an egalitarian society based on rejection of economic growth in favor of a smaller population eating lower on the food chain, consuming a lot less, and sharing a much lower level of resources much more equally. As Kathleen Courrier of the World Resources Institute, a premier environmental organization, put it in her review of John Young's *Sustaining the Earth,*

> environmentalists share an emerging analysis of the . . . economic order. The growing consensus among them . . . is that "the monopoly of wealth [in the Third World] by local elites and their demand for the consumer goods of industrial society have produced structural underdevelopment and political instability." . . .
> But here the consensus stops and the task facing postenvironmentalists begins to become clear. Most environmentalists are still pragmatists who believe that through mechanical and social engineering the bounty of growth can, in Young's words, "one day be enjoyed by everybody in a well-managed environment." Humanists, on the other hand, are wary of

the Trojan horse of postwar affluence. To them, economic growth is not worth the inequalities that have attended it. Environmental salvation seems possible only after the great gulf between the haves and the have-nots is closed.[1]

The idea is that "the redistribution of wealth can compensate for slowed economic growth." Such sentiments, in short, are not caricatures by opponents but an accurate rendition of what environmentalist-cum-postenvironmentalist leaders are trying to accomplish.

Their favored political mechanism would be an international treaty, modeled on the Montreal protocol that is designed to curb CFC emissions believed to be responsible for thinning the ozone layer.[2] Think (if you are persuaded you hold the truth) of the glory of it: no need to cope with regulations in different countries or, in America, in different states. Everything can be done uniformly and worldwide by central direction.

Suppose the warning of warming turns out to be false or vastly exaggerated. Well, as environmentalists are wont to say, it's only money. Besides, by then the greening, that is, deindustrialization of America will be well on its way.

In social science, a theory wins when it subsumes competing theories under its own categories. In society, a policy (or family of policies) prevails when it drives other policies in its wake. If preemption of other programs is a sign of a powerful policy, then a multitrillion-dollar carbon withdrawal policy would be the most powerful of all. For it will both preempt all new resources and (given a trillion-plus annual federal budget and a five- to six-trillion-dollar economy) a good part of present programs.

In other times, when one wished to stop something from being done, especially something with a scientific-technical component, it was usually necessary not only to muster political support but also to give something like plausible reasons. Nowadays, all that is necessary is to utter one or two cry-words—endangered species, spotted owl, wetlands, historical monument. New curses are heard: may your house or church become a historical monument or be designated a wetland, for then bureaucrats and judges will decide what you can do with what you used to think of as your own property. Indeed, it is difficult to imagine the American people agreeing to the pervasive and powerful regulation of activities, ranging from gas stations, body shops, photo stores, antifreeze, photocopy machine toner, storage tanks, on and on, except in the name of environmental protection and human health. But why do I mention this pattern of detailed regulation in an essay on global warming?

One has to admire the sheer nerve: advocate a policy of carbon withdrawal to delay the onset and reduce the severity of global warming, ward off arguments that much more needs to be known about the likelihood of its occurrence, the degree of warming expected, and the harm (even, possibly, help) that might result, by insisting there is no time to wait. Thus the proponents hope to subordinate the truth of what they are saying to the urgency of action: Act now! Now! Now! has been their constant refrain.

That this demand is ill-founded will become evident from Dr. Balling's judicious and probing account. While the evidence is being assessed, however, the advocates of carbon withdrawal rest their demand for immediate and massive action—the largest policy of all time with the possible exception of Social Security—on the ground that it represents a modest insurance policy. One does not often pay half one's income, no less national income, for insurance. In the meantime, however, before it becomes clear that modest warming would on balance be healthful, and that cooling is about as likely as warming, a vast regulatory apparatus will have been put in place. "Meantime" turns out to be at least half a century.

Yet if, as the science of the subject progresses (as Balling argues it is doing rapidly), we discover more about the likelihood and degree of global warming, won't we be able to shift back in the direction we were going? Not, I think, if the proponents of global warming prevail. For one thing, it will not be easy to get them to admit error. A skillful aspect of their presentation of policy is that there is no evidence, other than decades of freezing weather, that could, by their own criteria, prove them wrong. As things stand, their theory of warming is the program in their computer models. In urging us to accept these models, the proponents disarm their potential critics by saying that their programs lack the fine detail that would enable them to predict the weather in the short term or explain what is happening today in the world's regions or even account for past climate change. All the models are supposed to do, in their sponsors' view, is predict climate change for the earth as a whole decades ahead. Charming. But, as Balling points out, there is no contemporary or near-future evidence that could undermine the "religious" vision of an industrial capitalist civilization whose abuse of the earth has led Mother Nature to wreak a horrible vengeance, as if we were to burn in hell-like heat for our sins against the earth and her creatures.

Another requisite of the evidence-will-alter-policy thesis is that new findings be reported in the media and be treated with respect in Congress. Maybe. If the fate of the major report on acid rain is any guide, however, evidence to counter the warming thesis may well be ignored by the media or passed off as the work of prejudiced people who might (gasp!) have been contaminated by contact with industry.

Meanwhile, back in the mundane world of public policy, the threat of global warming will be used to justify nothing less than changing how we live. Private cars will give way to public transport. Every source of greenhouse gases will be regulated. The reduction of cattle herds may be justified on the grounds that their backsides emit too much methane; besides, they are too high on the food chain anyway. Then we will discover that carbon dioxide is neither the only nor the most important greenhouse gas by volume. Regulating methane, for instance, will involve curtailing rice production and other forms of agriculture. (When the regulators can reach termites I must leave what will be done to the reader's imagination.) Water vapor, by far the largest greenhouse gas, opens up limitless possibilities. In short, agriculture, industry, transportation, practically everything can be regulated with the aim of (or in the guise of) limiting global warming. One of many worries is that if the proponents have their way, by the time we know whether and to what extent global warming has occurred, the United States will have started down the path to deindustrialization, so that the knowledge and organizational capacity to cope with climate change will also have deteriorated. A closer look at the arguments, viewing the scientific positions through a cultural lens, will enable us to follow not only the "technics" but the "politics" of global warming.

GLOBAL WARMING OR GLOBAL COOLING?
A CLASH OF RIVAL MYTHS

Followers of scientist James Lovelock have embellished his notion (the Gaia hypothesis) that Mother Earth is akin to a human organism that knows better than the rest of us how to take care of herself. The suggestion that "nature knows best" appeals to environmentalists who wish human beings would tread lightly as if they were guests in the home of another greater than they. But the hint by critics of the warming hypothesis that the earth is like a thermostat calibrated to keep temperatures roughly even so that various processes operate as negative feedback to dampen fluctuations is unwelcome to advocates because it suggests that Mother Nature works contrary to their purposes. The point that Balling makes so well is that "maybe" is a poor basis for policy when the earth could as well be cooling as warming.

By beginning with these myths I do not mean to neglect scientific evidence but rather to help the reader structure the debate by placing the rival positions within these two models that weave stories of human dimension around these colossal goings-on. Myths, in my understanding, are models with a moral, which are not necessarily falsifiable.

ACTING BEFORE THINKING

The Associated Press reported that "A senior United Nations environmental official [Director of the New York office of the U.N. Environmental Program, UNEP] says entire nations could be wiped off the face of the earth by rising sea levels if the global warming trend is not reversed by the year 2000." Making this startling prediction in mid-1989, the UN official blamed burning of fossil fuels and the clearing of tropical rainforests for the greenhouse effect, under which polar icecaps would inundate coastal areas, swamping hundreds of millions of people.[3] Indeed, urging incoming U.S. President George Bush not to be "a political partisan who bends technical argument to support positions based on ideology," the National Academy of Sciences, together with the National Academy of Engineering and the Institute of Medicine, "among the nation's most prestigious scientific institutions," urged that global warming receive high national priority because "the future welfare of human society" is at risk. "Though there are still uncertainties," the NAS paper stated that "our current scientific understanding amply justifies these concerns."[4] In Canada, Minister of the Environment Lucien Bouchard, saying he cared nothing for his personal political survival, told reporters that "We must find a policy for this country to fight this terrible enemy, the subversive enemy, the silent enemy, which is destroying the planet now."[5] He argued "that the solutions are obvious. But there is no denying that enacting them will require paradigm shifts in human behavior."[6] Presumably, it is our behavior he wants to change.

Speaking for the environmental organization Greenpeace, Jeremy Leggett "answered the argument that it was better to wait for more scientific evidence before undertaking drastic solutions."[7] According to Leggett:

> This, then, is the heart of the matter. For organizations like Greenpeace, what comes first must be the needs of the environment—ahead of the expectation of multinationals to make billion dollar profits, ahead of the idea that freedom involves the right to pollute with impunity unless there is categorical evidence that you are swelling hospital waiting lists. The *modus operandi* we would like to see is: "Do not admit a substance unless you have proof that it will do no harm to the environment"—the precautionary principle. . . . The fact that proof of harm might come too late—or that proof is invariably hard to demonstrate with absolute certainty—only augments the license given the polluters.[8]

The general idea, as Eliot Marshall reported in *Science,* is that while all predictions about "global temperatures are riddled with uncertainty . . . it is not important to resolve the scientific issues before taking action."[9] As James R. Udall stated in the *Sierra,* the magazine of the Sierra Club, "In recent months the Beijer Institute in Stockholm, United Nations Environment Programme, World Meteorological Organization, Worldwatch Institute, World Resources Institute, Environmental Protection Agency, Sierra Club, and Woods Hole (Mass.) Research Center . . . independently concluded that the threat of global warming is so grave that action should be taken immediately."[10] Proof? Asking for proof only licenses the polluters! Those like myself who suppose that the greater danger comes from people who would control our behavior in the name of global warming, however, may want harder evidence. If so, Balling argues that we are unlikely to be rewarded.

IF THERE IS A SCIENTIFIC CONSENSUS, WHY ARE THERE CRITICS?

Essential for the existence of life forms on earth, greenhouse gases raise average temperature to about 60°F. It is also true that since the Industrial Revolution (and with increasing speed) carbon dioxide from coal burning and auto emissions as well as methane exuded by termites and cows will double sometime in the latter half of the next century.[11]

When we add in the rise in chlorofluorocarbons and nitrous oxides, in the last 100 years or so, these minor greenhouse gases have risen from 270 to approximately 350 parts per million of carbon dioxide with an extra 80 or so parts per million of the other gases. This increase and a projected doubling of CO_2 leads to predictions that average temperature would rise from 3°F to 9°F, far larger than the observed variation of plus or minus two degrees and sufficient at the higher ranges to create havoc on planet Earth.

With all this agreed, why don't scientists fall in line and lead the march for preventative measures to mitigate these potentially catastrophic effects? Some may believe the myth of Nature resilient, feeling that the earth will either adapt to or compensate for any large effects. Others may be wary of the immense cost, running to $3 to $4 trillion annually, to remove carbon dioxide and other gases from the environment.[12] The scientific critics of the global warming thesis are critical because their sense of the science and their readings of the findings, albeit incomplete, tell them that warming to the degree postulated is unlikely so that it would not be worthwhile to undertake huge, expensive, and disruptive measures to deal with a phenomenon that may never occur or, if it does, may occur much later and to a lesser degree than contemplated. Balling believes there is evidence that cooling is as or more likely than warming and therefore we ought not to take measures that will not only be immensely expensive, cutting our standard of living to abysmal levels, but also counterproductive for the very purpose of moderating the earth's weather. Without pretending that anyone has a patent on the truth, in view of the enormous press given to global warming, Balling indicates a number of the reasons for skepticism. In order to demonstrate that Balling's views deserve credence, I shall refer to other scientists who are similarly skeptical of the received wisdom.

ANOMALIES OF THE WARMING THESIS

In his protest against "the politicization of the global change issue," Patrick J. Michaels of the University of Virginia agrees with Balling's conclusions. Noting how the global warming thesis has swept public and political opinion despite its many weaknesses, Michaels calls attention to "remarkable inconsistencies":

The Northern Hemisphere, which should suffer less from oceanic thermal lag, is not warmer than it was 55 years ago. One study shows relative warming at night, which may be beneficial. The amount of global warming is at least a factor of two less than predicted by the most sophisticated models. If findings about urban contamination of climate records in the United States persists worldwide, the amount of warming is even less and may vanish to nearly zero. Polar regions, which should show amplified warming, have a temperature history during the past 40 years that is in fact opposite to what many would expect. Finally, there is evidence that other human-generated compounds may be mitigating the expected warming.[13]

Why, if global warming is a linear product of the accumulation of CO_2, Balling asks, did most of this century's warming take place before 1938 when less CO_2 was around? Between 1938 and 1970, the considerable decline in temperature led to predictions, from some of the same authorities who now predict warming, that a new ice age was at hand. And why, if as the earth warms the icecaps begin to melt, have temperatures at both poles been cooling slightly and icecaps showing some growth?[14] Let us pursue what has been happening in the earth's coldest regions as compared to what was supposed to have been happening.

"Glaciers in Norway have started to creep down from their mountain strongholds—growing bigger in apparent defiance of global warming." According to Professor Olav Orheim, the head of the Antarctic Section at the Norwegian Polar Research Institute, global warming theories have missed the side effects of higher temperatures, that is, more snow.[15] Antarctica, it turns out, has 90 percent of the world's ice. Whereas in 1985, a committee of the National Research Council chaired by Mark F. Meier of the University of Colorado at Boulder predicted that a 3°F increase in average global temperature would lead to a rise in sea level of about one meter, this occurring around 2100 when the carbon dioxide in the atmosphere doubled its 1950 value, new information suggested the rise would be only one-third of a meter. Nor would the West Antarctic ice sheet disintegrate. Instead, it would likely grow, thereby pulling water out of the oceans. What is more, to strike a note that will keep reappearing, scientists do not understand the general reasons why the level of the sea has been rising in recent times. Their theories fall far short of the observed 1.3 millimeters annually. "This means," Meier told a reporter for *Science News,* that "our understanding of the system is not very good at the moment."[16] The account by Balling that follows provides chapter and verse to this proposition.

"I was shocked to discover, when I went to Oak Ridge in 1975," Freeman J. Dyson recalled, "that nobody knew what happened to half the carbon that we were burning." Roughly 6 gigatons of carbon (six thousand million metric tons) is being poured into the atmosphere, but something less than half, 43 percent, is somewhere else, though nobody knows where. "This," Dyson continues, "is *the mystery of the missing carbon* [emphasis added]. It is preposterous to claim any ability to predict the future of the carbon cycle so long as we lack a rudimentary understanding of what has happened to the carbon in the past."[17] Is it in trees or plants or oceans? In kelp beds? Where? As a scientist should, Balling patiently shows that our knowledge of key variables is so small that our first task, long before we throw the nation and the world into turmoil, is to make up for our ignorance.

IF OUR FUTURE IS CLOUDY, IS THAT BECAUSE MODELS IGNORE THEM?

There is general agreement, so far as climatologists like Balling know, that *both the heating and cooling effects of clouds dwarf by five to ten times whatever mankind does to increase global warming.* Relatively modest changes in how clouds are modeled—whether they are considered at all or how they are considered, that is, what kinds of clouds under what conditions—can make significant differences in the amount of cooling or warming predicted, differences more than large enough to cancel or reverse or exacerbate global warming. A comparison of 11 greenhouse models conducted by Robert D. Cess of the State University of New York at Stony Brook reveals that assumptions about how clouds work led to differences in predictions of warming by three times over. The first step has been taken by V. Ramanathan of the University of Chicago who demonstrates, for the first time, amazingly enough, using observations by two satellites so as to cover a large part of the earth, that clouds actually cool the earth somewhat. As one of the American global climate modelers, Michael Schlesinger of University of Illinois, says, "You have every right to be very, very skeptical of the results. But this is the best that we're doing."[18] This best, Balling is at pains to show, is far from good enough.

For good reason, J. K. Angell tells us, aficionados call clouds "the wild card in the game of global [climate] change."[19] The trouble is, as Balling informs us, some cloud formations cool while others warm. Because cirrus clouds let much of the sun and its ultraviolet light through, they warm; because they are formed in the highest and therefore coldest levels of the troposphere, the infrared energy they absorb is not radiated away but stays within them, thus exciting its molecules and adding to the warming. Marine stratocumulus clouds, by contrast, cool the atmosphere. They are thin; thus the droplets of water that fill them radiate deep into outer space, thereby cooling the earth. Since cirrus clouds cover only 16 percent of the earth, whereas stratocumulus clouds cover around 34 percent, their cooling effect dominates. Anyone who has tried to deal with difficult problems can sympathize with modeler Anthony Slingo of the National Center for Atmospheric Research, who thinks that "The trouble is the lack of a very firm theoretical foundation for the way we treat clouds in a general circulation [climate] model. When we get a result on climate change, we don't really know what confidence to give it."[20] To which, in effect, Balling says, Amen.

How much do clouds, or rather their treatment in models, matter? A change in the way clouds were modeled by adjusting the amount of water they contained led modelers of the British Meteorological Office to reduce their 5°C warning after CO_2 doubled to 2°C.[21] "We're not dealing with an exact science here," said modeler Tony Slingo as he expected estimates of warming to go up and down like yo-yos.[22] What Balling is trying to tell us is that this state of affairs may be all right for scientists trying out pet theories, but not for society. And I must confess that the yo-yo theory of public policy does not much please me either.

The layman (this writer) assumes without thinking that the earth gets cooler or warmer. Foolish thought. How could one be so unsophisticated? It turns out, based on research by Thomas Karl at the government's Climate Data Center in Asheville, North Carolina, that there has been a modest *rise in nighttime temperatures while daytime temperatures have dropped* a little. Obviously (or should one omit such a term?) nighttime warming would be easier to bear and have less deleterious effects on heat-sensitive people and plants than daytime warming.[23]

It is known that the oceans serve as an enormous carbon sink, but it is not known how much and to what extent the oceans absorb either carbon or, significantly, heat. It is not only clouds, therefore, but oceans that have been poorly portrayed in existing models.[24] Thus, when the National Center for Atmospheric Research observed the influence of oceans, it reduced its predictions of warming contingent on a doubling of CO_2 from 7°F to 3°F. Such considerations led Robert Cess to state that "We don't know what these models are doing. As presently formulated [they cannot be used to predict future global warming] and whether they can ever be used for that purpose is problematical."[25] A major effect of Balling's patient argument to the same end, I hope, will be to show that there is no point in buying insurance against a contingency whose probability and extent you cannot predict.

IF WE DON'T KNOW THE TEMPERATURE, JUST WHAT DO WE KNOW?

With all these unknown or insufficiently known variables, let alone their interactive effects, such as convection on clouds and clouds on oceans, experts as well as laymen are puzzled. At least, the ever-hopeful layman may think, we know the temperature. Not so. Because most temperature readings in the past have been taken in cities, and there is reason to believe that cities are heat islands, temperatures may be over-estimated.[26] Old measurements on oceans, for instance, and newer ones beginning to come in from recent efforts, so far suggest that the oceans have not been warming, at least not nearly enough to support the linear carbon dioxide thesis.[27] And there are also efforts underway to measure temperatures in the stratosphere, presumably unaffected by the heat generated by the passions below, to see where they fit.[28] Whether one looks on land or sea or in the air, therefore, temperatures, including the instruments to measure them, have become controversial.[29] I read Balling as saying that there is no point in telling people to jump to safety if you cannot tell whether they will fly up or drop down.

ADVANTAGES OF WARMING

Another way to counter an argument whose implications you do not like is to show that they are advantageous rather than disastrous. This Balling does in exemplary fashion. Without admitting that global warming will occur, at least to the extent predicted by proponents, opponents create the picture of the benign movement from a carbon-starved to a carbon-rich environment in which carbon serves as an immense fertilizer that not only vastly increases plant and tree growth but also and simultaneously (How wonderful nature can be!) sharply reduces the need for water. Global warming at, say, 2°C to 3°C, according to the Intergovernmental Panel on Climate Change meeting in Berkeley Springs, West Virginia, benefits increased food production and water availability as well as forestry area and volume, thereby dwarfing the costs.[30]

WILL THE EARTH TAKE CARE OF ITSELF?

In a report issued by the George C. Marshall Institute called "Scientific Perspectives on the Greenhouse Problem," Frederick Seitz, a past president of the National Academy of Sciences and of Rockefeller University, Robert Jastrow, founder of the Goddard Institute for Space Studies, and William A. Nierenberg, director emeritus of the Scripps Institute of Oceanography at UC San Diego (no kids these) argued that global modeling was so unsatisfactory that any number of other explanations could do just as well as the CO_2 hypothesis. Among these alternatives for creating the warming trend already observed was increased activity by the sun whose variations, they argue, come closer to the warming trend than do the accumulation of greenhouse gases. In response, Stephen Schneider argues that the sun or any number of other factors could explain past warming, but that would not affect what the rapid buildup of greenhouse gases could do in the future.

The Marshall Institute paper went further to suggest that solar activity might well decrease in the next century, thus leading to a smaller version of the ice age, thereby offsetting any greenhouse effect. In response, John Eddy of the University Corporation for Atmospheric Research says what opponents of solar warming usually say to its proponents, namely, that no one can predict future solar activity. But the reader should take the point: Seitz, Jastrow, and Nierenberg want to derail the global warming express by showing that alternative hypotheses, albeit just as speculative, do equally well or better. Thus they recommend $100 million for supercomputers to make the research go faster and say there is time to wait, exactly what the proponents deny.[31]

The best known scientific critic, Richard Lindzen, Sloan Professor of Meteorology at the Massachusetts Institute of Technology, a member of the National Academy of Sciences, and, most importantly, a major contributor to current research on atmospheric behavior, has denied the likelihood of substantial global warming. In part, his objection goes to the declaration by the National Academy of Sciences that there was a consensus on global warming when he knew that he and many of his colleagues around the country did not agree. Asked, in a questionnaire from NAS, whether carbon was accumulating and whether, if unhindered and no other factor was present, this would lead to warming, Lindzen agreed, as would every scientist. Global warming, he says, "is the only subject in atmospheric science where a consensus view has been declared before the research has hardly begun." That is a concise statement of the theme of this book.

Essentially, Lindzen postulates that negative feedback will take care of what might otherwise have been global warming. And he provides a list, longer than the one here, of important factors global models leave out. In the models he disputes, warming creates more water vapor that always acts with positive feedback to trap more heat and warm up the climate. To Lindzen, the opposite is more plausible. In his model, the earth doesn't need all those well-wishers. His sense of the science is that Mother Earth will protect herself.

"Does he [Lindzen] have a calculation," opponents like Stephen Schneider ask rhetorically, "or is his brain better than our models?" Schneider says that skeptics like Lindzen "count all the negative feedbacks we don't know about, and forget about the positive ones we don't know about."[32] Nonetheless, Lindzen's prestige is sufficiently great that a petition was circulated among what *Science* appropriately terms "700 heavy hitters" including half the members of NAS and 49 Nobel Prize winners calling attention to the dangers of global warming. The petition was circulated by the Union of Concerned Scientists.[33] Consider in this connection a comment made by John Perry, who is a meteorologist and is staff director of the Board of Atmospheric Sciences and Climate at the National Research Council, in countering the Marshall Institute report: "If the report had just said, in an evenhanded way, 'Don't rely on the models because there are hellacious uncertainties,' we all would have applauded."[34] Were there in fact common agreement on this position, and were it widely circulated and repeated among politicians, who would support a $3 to $4 trillion effort to reduce so uncertain an effect when a few years of intensive effort should reveal a great deal more?

A number of the leading critics have given reasons, recounted here, for doubting global warming. Perhaps because they regard the cumulative effect of their objections to be decisive, they do not list them all in the same place, or go into the reasoning for each one, or connect the different parts of their arguments to each other. Balling does. His work is not only accessible to the layman, but it also has the merit of being the most comprehensive account available. Advocates of carbon withdrawal are afraid of what warming will do to the earth, as I am afraid of what their policies will do to us. Thus I doubly welcome Balling's calm marshalling of evidence.

Suppose that someone wished to argue that global freezing would be more likely to occur in the future than global warming. One could then take attention away from carbon dioxide and go to sulfur dioxide that comes in significant measure from coal-burning plants, just like CO_2. But SO_2 has a cooling effect because its tiny bright particles reflect heat out of the atmosphere. Thus critics of global warming say that if the proponents perturb their models with SO_2 instead of CO_2 they could get cooling instead of warming.[35] Balling's chapter, so far as I know, represents the first time this scattered evidence has been put together.

If the evidence of global warming is weak, as it is, and the ability of climate models to predict the near-term let alone the long-term future is gravely in doubt, all of which should now be evident, why go into a panic? When one reads headlines like, "Environmentalists hope for scorcher: Aim is to avert governmental complacency on 'greenhouse effect',"[36] I wonder whether incantation has become a substitute for understanding.

More forthcoming than most, in an interview in *Discover* magazine, Stephen Schneider stated that

> scientists should consider stretching the truth to get some broad-base support, to capture the public's imagination. That, of course, entails getting loads of media coverage. So we have to offer up scary scenarios, make simplified, dramatic statements, and make little mention about any doubts we might have. . . . Each of us has to decide what the right balance is between being effective and being honest.[37]

If everyone is trying to counter someone else's outrageous comments, it will not be possible to figure out even approximately where the truth, insofar as it is known, lies. (I am reminded of the difficulty the Soviet Union government, before its recent demise, had in reconstructing economic, health, and budgetary figures because various people in the past altered them for political purposes without standardizing these alterations.) Righting environmental wrongs, if that is the motivation, is not well done by environmental ignorance. I would like to align myself with a heartfelt comment by biologist Michael Gough in a letter to *Science* concerning the efforts of veterans' groups and families to claim that soldiers in Vietnam got cancer because they were exposed to Agent Orange, a letter written before more conclusive evidence appeared that low-level, intermittent exposures to dioxins could not cause cancer in human beings. Gough writes that

> Agent Orange is one of the last vestiges of the nation's torment over the Vietnam War. Many members of Congress as well as many citizens are ashamed of our treatment of Vietnam veterans during and immediately after the war, a feeling that I share. But that guilt also fuels the continued search for evidence that Agent Orange "did" something to the health of veterans. It is ironic that the mental and emotional anguish caused by all wars is largely ignored while we search in vain for a chemical cause for diseases that occur as frequently in nonveterans as in veterans, and, so far as can be told, as frequently in veterans not exposed to Agent Orange as in those who were exposed. This is not the way to right any wrongs that may have been done.[38]

Exaggeration or worse for the right cause is the wrong way.

PUBLIC POLICY

As for the presumed economic effects of a doubling of carbon dioxide, the latest word comes from Yale economist William Nordhaus. Where some activities like cardiovascular surgery and the fabrication of microprocessors are likely to be unaffected by climate change, given that they are already in controlled environments, other activities like agriculture are sensitive to climate change. His best estimate of the effects on agriculture, given that CO_2 is a powerful fertilizer, is plus or minus $10 billion. The additional costs of protection in land areas, assuming that the projected rise in the sea level is not larger than that experienced in the last hundred years along the Gulf Coast, is not expected to be much greater than it is now. In regard to energy, the demand for air conditioning will surely rise but would be offset by the decrease in demand for space heating. Forest products should benefit somewhat; so should recreation as warmer areas expand with water skiing going up and snow skiing coming down. "We estimate," Nordhaus concludes, "that the net economic damage from the canonical 3°C warming, in terms of those variables that have been quantified, is likely to be around 1/4 of one percent of national income," with an error bound of 2 percent of Gross National Product either way.[39] The implication is that spending a third to a half of GNP a year to remove carbon from the earth in order to achieve a saving of one-quarter of one percent or, given the worst case, two and a quarter percent, would not be wise. Hence, Nordhaus suggests calibrating what is done to the evidence as it appears about whether global warming is taking place and at which magnitude.[40]

After arguing that a variety of objections still leaves the thesis of global warming plausible, especially as warmer years succeed each other, modeler Stephen H. Schneider turns to the question as to whether immediate and drastic action is desirable. He quotes Andrew A. Solow of Woods Hole, who rejects the arguments in favor of acting "before it is too late" on the grounds that it "applies equally to an invasion by aliens from space. More seriously, this argument neglects the costs of over-reaction."[41] This, Schneider says to his credit, "is a legitimate issue but so too is the cost of under-reaction." He feels "that we insult the environment at a faster rate than we can predict the consequences." If we wait for greater certainty from scientific findings, "The decision will not be cost free. It will be bought at the price of forcing us and other living things to adapt to a much larger dose of change than if we were to act today." He wants to act now because "circumstantial evidence suggests that climate change over the next several decades could proceed at unprecedented rates. But again, this cannot be demonstrated beyond a reasonable doubt without another decade or two of measurement and research."[42]

I think that doubts about the accuracy of global models, coupled with Nordhaus's demonstration that effects against which we would be guarding are likely to be far smaller than the damage done by the expenditure involved, make it desirable to learn more before acting. But I do not suppose that this matter is cut and dried. Far from it. I hope to have been successful in showing that reasonable people could well believe that "the end of industrial society as we know it," which is what would be involved in a policy of carbon withdrawal, is not dictated by the preponderance of evidence.

The greenhouse effect (or, indeed, any substantial effect whatsoever) has to fit into a larger set of understandings; if knowledge about them is nonexistent or, more likely, weak, then so will be predictions of the effects of a single factor. When scientists speak of fundamental science of a subject or, more casually, of what is known, they refer both to a body of propositions about their subject, with some evidence to support them, and to intuitions based on immersion in the subject as well as common understandings with other scientists. When anomalies arise that are inexplicable according to generally accepted propositions, scientists are thrown back on their common understandings. When these are ruptured, rival myths, based on fact but beyond fact, contend for supremacy. Thus one can sum up the debate over global warming this way: Proponents: positive feedback (or Murphy's law, everything that can go wrong, will go wrong). Opponents: negative feedback (natural systems will act to dampen fluctuations).

If readers come away from this book wondering whether the debate over global warming is just a disguise for the same old ideological conflicts that divide Americans politically—central planning versus free enterprise, regulation versus free enterprise, spontaneity versus control—they have a point. I suspect that the global modelers who call for regulation of carbon emissions are more liberal and democratic than their critics. But that is not all there is to it. There is also a common respect for evidence and a common commitment to scientific methods. In the interest of nurturing respect for evidence, we should make all possible use of what we share and hold in common. There are, as we have seen, numerous ways of sidestepping evidence, but it still matters. If it did not matter at all, there would be no argument. Proponents and opponents would not bother to try to persuade one another with scientific claims. Just as hypocrisy is a hostage to virtue, so we notice that, however small the respect for evidence appears to be, public discourse is carried on in the language of science, not of nature mysticism. Those who want to understand global warming as a fascinating problem in applied science should read Dr. Balling's learned account.

The implications for public policy of this book are even more powerful than Balling, in his characteristically cautious way, states them. If the temperature record varies over time and space, day and night, city and country, and if a delay of a decade or two will have virtually no effect on the extent of warming, should it occur, and if modest amounts of warming are likely to be beneficial, then it would be foolish to act now. Committing the United States to drastic carbon withdrawal now would go against preponderant evidence. Researching rather than reducing global warming is not just a plea to gain time; researching is the only way to gain knowledge needed for public policy. Balling shows that if we acted on what we thought we knew five years ago on sulfates, we would probably have increased warming.

The Alabama survey of professional people Balling discusses in his last chapter is instructive. The temperature record shows cooling; the professionals believe there has been warming and that carbon dioxide is responsible. Apparently, they learn what to believe by reading the newspapers or watching television, for their views are consistent with portrayals of warming in the media, albeit not with the evidence. All of us, outside our narrow specialties, are laymen in other fields.

Even if significant warming were to occur, Balling notes, the policy implications drawn by advocates are likely to be wrong-headed. They call for reduction of carbon emissions by advanced industrial nations like the United States. But these are the efficient nations that produce far more output per unit of energy (and hence emissions) than do nations with less productive economies. The vast pollution in the former communist countries of Eastern Europe and what was once the Soviet Union speak eloquently of the importance of efficiency. The sheer sanity of Balling's "One way to reduce CO_2 emissions substantially is to make much of the world as 'efficient' as the United States" is hard to beat.

In conclusion, I would like to second Balling's belief that ordinary intelligent people can read his book and come to their own considered judgment. There is no better place to begin than with his careful, informative, learned, and fair account of the scientific issues surrounding global warming. His conclusions are hard to resist because they are so sound and sane: watchful waiting for research results, yes; impulsive and massive action, no.

Suppose the alternatives are, as Balling calls one of them, "business as usual," and what I have termed "carbon withdrawal." These are antiseptic terms. The business-as-usual scenario implies the maintenance of industrial civilization with its high standard of living, a living standard that generates knowledge to cope with whatever danger may arise. The carbon withdrawal scenario, depending on how extreme, would drive us back either to the beginnings of the Industrial

Revolution or to the 1930s with its much lower standard of living and level of scientific understanding. If the advocates of carbon withdrawal are on the right track in believing that global warming will occur in severe degree, Americans will be poor but healthier. If Balling is on the right track in urging caution, as I believe, and warming will not occur or be rather mild, Americans would be both poorer for having spent huge sums and sicker because health declines with a lower standard of living.

If, going beyond what Balling writes to my own views, the motive that leads to resolving all uncertainties in favor of a policy of carbon withdrawal is to equalize differences among people, then the advocates of global warming may have exactly the right policy. Whether one is willing to accept considerable inequality providing that the standard of living of all peoples rises or whether one prefers much greater equality of condition even if the standard of living of all people falls is one of the diagnostic questions of our time.

NOTES

1. Kathleen Courrier, "Postenvironmentalism," *Issues in Science and Technology,* vol. 7, no. 4 (Summer 1991), pp. 102–103.
2. Whether the protocol is desirable depends on whether there is actually a hole (or thinning) in the ozone layer. My preliminary assessment is "Maybe, but by no means certainly." As knowledge grows, the initial verdict may well be overturned.
3. Associated Press, "Warming alert is sounded," June 30, 1989.
4. Philip Shabecoff, "Bush is urged to fight threat of global warming: The National Academy of Sciences says society is at risk," *New York Times,* January 5, _____ .
5. "Earth's fate my priority, Bouchard tells committee," *The Globe and Mail,* Toronto, October 27, 1989.
6. W. Booth, "Action urged against global warming: Scientists appeal for curbs on gases," *Washington Post,* February 2, 1990.
7. Jeremy Leggett, *Global Warming: A Greenpeace View* (Oxford/ New York: Oxford University Press, 1990). See also J. Leggett, "The new Manhattan Project: America's weapons labs are turning to a new enemy," *New Scientist,* September 16, 1989.
8. Leggett, *Global Warming: A Greenpeace View,* p. 459. The Greenpeace "precautionary principle" is the same as the one I discuss in my *Searching For Safety* (New Brunswick, N.J.: Transaction Press, 1988) as "No trials without prior guarantees against error." This note of

immense alarm together with the reiterated demand that action be taken before it is too late to prevent catastrophe (act now!) is the leitmotif of environmentalists' demands. Thus the executive director of the Union of Concerned Scientists, Howard Ris, likened the greenhouse effect to the problem of someone who is locked in a garage with the motor running and no keys and, eventually, of course, no air. "Only by taking action now," Ris said, "can we insure that future generations will not be at risk." Eileen Klineman, "Scientists: U.S. must act swiftly," *Press Democrat* (Santa Rosa, California, newspaper), April 20, 1990, pp. B1–2.

9. Eliot Marshall, "EPA's plan for cooling the global greenhouse," *Science,* vol. 243 (March 24, 1989), p. 1545.

10. James R. Udall, "Climate shock: Turning down the heat," *Sierra,* July/August 1989, p. 28.

11. Michael Precker, "Blame the cows? Really?" *San Francisco Examiner,* November 25, 1990.

12. *New York Times,* November 19, 1989; Speech by EPA Director William Reilly to World Affairs Council, June 27, 1989, p. 13.

13. Patrick J. Michaels, "Crisis in politics of climate change looms on horizon," *Forum,* vol. 4, no. 4 (Winter 1989), pp. 14–23.

14. H. W. Ellsaesser, "A different view of the climatic effect of CO_2—Updated," *Atmosfera,* vol. 3 (1990), pp. 3–29.

15. Alice Dierdoyle (Reuters), "In Norway, glaciers are growing bigger," *Los Angeles Times,* October 28, 1990.

16. Richard Monastersky, "Predictions drop for future sea-level rise," *Science News,* December 16, 1989, p. 397.

17. Freeman J. Dyson, Institute for Advanced Study, Princeton University, "Carbon dioxide in the atmosphere and the biosphere," Radcliffe Lecture given at Green College, Oxford, October 11, 1990. The substance of this claim was published by Dyson as, "Can we control the carbon dioxide in the atmosphere?" *Energy,* vol. 2 (1977), pp. 287–291.

18. Richard A. Kerr, "How to fix the clouds and greenhouse models," *Science,* vol. 243 (January 6, 1989), pp. 28–29.

19. J. K. Angell, "Annual and seasonal global temperature changes in troposphere and low stratosphere," *Monthly Weather Review,* no. 107 (1986), pp. 1922–1930.

20. Richard Monastersky, "Cloudy concerns: Will clouds prevent or promote a drastic global warming?" *Science News,* August 12, 1989, pp. 106–107, 110. I cannot here indicate all or even most of the differences. One modeler, for instance, argued that sulfur emissions from plankton would increase the formation of clouds that reflected back heat thereby increasing global warming. Not long after another modeler observed that the emissions of sulfur from human activities were roughly

double that of plankton. See the article by John Horgan, "Pinning down clouds," *Scientific American,* vol. 260, no. 4 (May 1989), p. 24.

21. William A. Nierenberg, "Atmospheric CO_2: Causes, effects, and options," *Chemical Engineering Progress* (August 1989), pp. 27–36; Richard Monastersky, "Looking for Mr. Greenhouse," *Science News,* vol. 135, no. 14 (April 8, 1989), pp. 216–217, 221; Richard Monastersky, "Global change: The scientific challenge," *Science News,* vol. 135, no. 15 (April 15, 1989), pp. 232–235.

22. William Booth, "Climate study halves estimate of global warming," *Washington Post,* September 14, 1989; J. F. B. Mitchell, C. A. Semar, W. J. Ingram, "CO_2 and climate: A missing feedback," *Nature,* vol. 341, no. 6238 (September 14, 1989), pp. 132–134.

23. Booth, "Climate study halves estimate of global warming."

24. R. A. Newell et al., "Long-term global sea surface temperature fluctuations and their possible influence on atmospheric CO_2 concentration," *Pure and Applied Geophysics,* vol. 116 (1978), pp. 351–371.

25. William K. Stevens, "Skeptics are challenging dire 'greenhouse' views," *New York Times* National Edition, December 13, 1989.

26. T. R. Karl and P. D. Jones, "Urban bias in area-average surface air temperature trends," *Bulletin of the American Meteorological Society,* vol. 70 (1988), pp. 265–270; P. J. Michaels et al., "Anthropogenic warming in North Alaska?" *Journal of Climatology,* vol. 1 (1988), pp. 942–945.

27. George C. Marshall Institute, *Scientific Perspectives on the Greenhouse Problem* (Washington, D.C., 1989), p. 5; *Global Ocean Surface Temperature Atlas* (Bracknell, England: Meteorological Office, 1990).

28. R. Spencer and J. Christy, "Precise monitoring of global temperature trends from satellites," *Science,* vol. 247 (1990), pp. 1558–1562.

29. William K. Stevens, "Skeptics are challenging dire 'greenhouse' views," *New York Times,* December 13, 1989, pp. A1, B12; D. Rind, E. W. Chiou, W. Chu, J. Larsen, S. Oltmans, J. Lerner, M. P. McCormick, and L. McMaster, "Positive water vapour feedback in climate models confirmed by satellite data," *Nature,* vol. 349, no. 6309 (February 7, 1991), pp. 500–503; William E. Reifsnyder, "A tale of ten fallacies: The skeptical enquirer's view of the carbon dioxide/climate controversy," *Agricultural and Forest Meteorology,* vol. 47 (1989), pp. 349–371; Gilbert W. Bassett, Jr., "Breaking recent global temperature records," typescript, July 1991.

30. Intergovernmental Panel on Climate Change, "Climate impact response functions," Report, September 11–14, 1989; see also the discussion in Warren Brookes, "Global warming benefits?" *Washington Times,* March 12, 1990; Robert W. Carlson et al., "Photosynthetic and growth

response to fumigation with SO_2 and elevated CO_2," *Oecologia,* vol. 54 (1982), pp. 50–54. To oppose the view that all good things go together, see Janet Raloff, "Not all plants will thrive in 'greenhouse'," *Science News,* vol. 136, August 12, 1989, p. 134.

31.   See the interesting interplay in Leslie Roberts, "Global warming: Blaming the sun," *Science,* vol. 246 (November 24, 1989), pp. 992–993.

32.   Richard A. Kerr, "Greenhouse skeptic out in the cold," *Science,* vol. 246 (December 1, 1989), pp. 118–119.

33.   Constance Holden, "Global warming petition," *Science,* February 23, 1990, p. 919.

34.   Leslie Roberts, "Global warming: Blaming the sun."

35.   Warren T. Brookes, "The global warming panic," *Forbes Magazine,* vol. 144, no. 14 (December 25, 1989), pp. 96–102.

36.   William Booth, *Washington Post,* June 21, 1989.

37.   *Discover,* p. 47.

38.   Michael Gough, Letter, *Science,* vol. 245 (September 8, 1989), p. 1031, then with Resource for the Future, is now with the congressional Office of Technology Assessment.

39.   William D. Nordhaus, "Slowing the greenhouse express: Economic policy in the face of global warming," in Henry Aaron, ed., *Setting National Priorities* (Washington, D.C.: Brookings Institution, 1990), pp. 185–212.

40.   Nordhaus, "Slowing the greenhouse express."

41.   Quoted by Schneider from Andrew A. Solow, "Pseudo-scientific hot air: The data on climate are inconclusive," *New York Times,* December 28, 1988; and his "Greenhouse effect: Hot air in lieu of evidence," *International Herald Tribune,* December 29, 1988.

42.   Stephen Schneider, "Global warming: Is it real?" pp. 24–30. See also Schneider, *Global Warming: Are We Entering the Greenhouse Century?* (San Francisco: Sierra Club Books, 1990).

1

INTRODUCTION TO THE GREENHOUSE EFFECT

From time to time, scientific questions surface that capture the imagination of researchers from many disciplines, the attention of decisionmakers, and the interest of the general public. The planet may be threatened by these matters (e.g., acid rain, nuclear winter), or they may promise some enormous benefit for the inhabitants of the globe (e.g., cold fusion). The demand for experts in related fields increases, new "experts" emerge in the public eye, and a number of positive feedback mechanisms fuel the interest and perceived importance of the original scientific question. In the late 1980s and early 1990s, the greenhouse effect followed this pattern and emerged as possibly the most important scientific question of our time. The greenhouse effect is front-page news, the cover story for many popular news magazines, and the focus of dozens of television documentaries. The storyline is clear and simple—greenhouse gases are increasing in concentration and their many effects are threatening the planet. We must act now to avoid disaster!

Such important and popular scientific issues often lead to fiery debates among a cast of characters with assorted underlying political, social, and economic motives. The greenhouse issue has not escaped this fate. And while the public has been bombarded with the apocalyptic view of the greenhouse world, the professional literature is alive with debate about the likely climatic outcomes of increasing the atmospheric

1

concentration of the many greenhouse gases. The threat of a global calamity makes news, but this catastrophic view is under serious attack by many professional climate scientists.

Accordingly, the purpose of this book is to provide the reader with some background to the greenhouse issue, present an analysis of the certainties and uncertainties for future climate change, and examine the most probable changes in climate that may occur as the greenhouse gases increase in concentration. These gases increased substantially over the past century, and a recognizable and internally consistent pattern of climate change has emerged already. These observed changes, coupled with an appreciation of the theoretical predictions of the global climate models, provide considerable guidance for the most probable climate changes that will take place in the years to come. Developing an understanding of these changes may prove valuable to the decision-makers who must consider various policy options to cope with the perceived threat of global warming.

THE "POPULAR VISION"

The greenhouse effect typically is presented as a series of interrelated climate changes and ecological disasters that stem from anthropogenic (human-caused) emissions of various greenhouse gases. This "popular vision" (Michaels, 1990) of the greenhouse world sees little or no benefit in the resultant changes, but rather tends to view the greenhouse effect as a great debacle. According to this highly popularized viewpoint, the greenhouse gases increase in concentration, they trap increasing amounts of the earth's heat energy, and the planetary temperature rises quickly by 4°C (7°F) or more. At various stages of this transformation, an assortment of interrelated changes devastate the ecosystem of the planet.

According to what I once referred to as the "greenhouse gospel" (Balling, 1990), the greenhouse-induced increase in heat raises the rate of evaporation of water from the land surfaces. Even if the rainfall increases slightly as a consequence, the surge in evaporation (and transpiration of water from plants) completely outstrips any rise in precipitation, and aridity in many parts of the globe increases. The popularized view of the greenhouse effect inevitably contains a very strong image of increased droughts, particularly within the agricultural heartlands of North America and Europe. The Dust Bowl returns, and the highly specialized agricultural systems are unable to cope.

While the problems compound in the agricultural regions, the coastal areas appear to be in even greater peril. Again, according to the "popular vision," the rapid warming of the earth melts substantial portions of the planet's icecaps and glaciers. As this ice melts, the highly reflective snow and ice surfaces are lost, and the less reflective, underlying terrestrial surfaces are exposed. Even the melting of sea ice reveals the much darker ocean surface. More of the sun's energy is absorbed by these surfaces once covered by snow and ice, local temperatures increase, and more snow and ice is melted (Huybrechts et al., 1991). The huge quantity of water tied up in the ice is released, and the sea level rises. Even more important, the oceans continue to warm and undergo thermal expansion thereby furthering the rise in water levels. The low-lying cities along the ocean coastlines are flooded, unprecedented migrations of human populations are triggered, and coastal ecosystems are destroyed. Maps of the continents need revision as the landmasses of the earth are swallowed by the rising waters.

Along with the higher temperatures, melting icecaps, sea-level rises, and increased frequency and severity of droughts, the "greenhouse gospel" predicts an increase in severe storm activity. The tropical seas warm, these ocean surfaces fuel larger and more intense hurricanes, and coastal areas in many parts of the globe are exposed to calamities as supersized hurricanes (labeled hypercanes) roar over the horizon. Obviously, the large hurricanes, combined with the rising sea level, will generate enormous storm surges that crush any development in these coastal areas.

Inland ecosystems are not spared in this apocalyptic and pessimistic view. The increased stress caused by declining soil moisture weakens forests and grasslands and increases the potential for large-scale wildfires. Indeed, the image of large fires in the western United States has become an additional and integral part of the popularized view of the greenhouse world. Streams and rivers dry up, selected pests thrive, new diseases appear, agricultural systems fail, economies of the world are shocked, widespread human migrations occur, and in general, the greenhouse apocalypse is realized.

The greenhouse catastrophe occurs within our lifetime or the lifetime of our children. And to make matters even worse, the long atmospheric lifetime of many of the greenhouse gases, along with the long period of time required for the oceans ever to cool back down, produces an almost irreversible set of deleterious consequences. Immediate action is needed, and a host of social and economic measures have been proposed to halt this ominous threat. We have, according to the "popular vision," the potential for great disaster, and luckily, we have the ability through our combined actions to reduce the severity of the greenhouse effect. The developed nations must live with less, developing nations must never make the environmental mistakes of their developed neighbors, fossil fuels must be viewed as greenhouse villains, and of course, deforestation must be reversed. In general, we must restructure our social, political, and economic activities or someday soon, we (or even worse, our children) will be forced to pay the greenhouse piper.

Is the "popular vision" nothing but environmental hype? Does it have any scientific basis? Is it vastly different from the viewpoint shared by professional climatologists? And most important, is the "popular vision" correct; are we moving ever closer to an unimaginable global-scale ecological catastrophe? And if we are, can we really do something to stop the dreaded global warming?

As we shall see throughout this book, the "greenhouse gospel" does in fact contain much obvious exaggeration that should not be taken so literally. Many of its predictions are seriously flawed and lack much scientific basis. Other predictions are well grounded in solid science and can withstand any number of theoretical and empirical tests. We will find that most greenhouse effects will be rather moderate and are not likely to produce ecological chaos. We will also find that more than a few probable greenhouse-related changes may even prove to be beneficial to the natural environment and the human inhabitants of the planet. Furthermore, our actions to cope with the perceived threat are likely to have only moderate impacts on the buildup of the greenhouse gases. We will put our emotions aside and carefully examine the very notion that we can all act now and stop global warming.

THE SCIENTIFIC BASIS FOR THE "POPULAR VISION"

The origins of the "popular vision" are relatively easy to trace over the past few decades. Up to the mid-1970s, climatology was a relatively minor component of the atmospheric sciences, many of its practitioners interested in climate classification and the routine management of climatic data. Even in the mid-1970s, few climatologists were trained in atmospheric physics, and very few were developing numerical climate models. Many leading climatologists were working on applied problems linking climate to food production, water supplies, and energy demands. Many of today's major scientific journals in climatology did not exist, and few young scholars were attracted to the field. However, in the mid to late 1970s, the United States, as well as many other parts of the globe, experienced a string of bitterly cold winters that had a profound impact on the field of climatology.

Rather suddenly in the 1970s, a host of climatologists appeared in the public eye armed with long-term climate records, a few results from crude numerical climate models, and a threatening tale. Their work had allowed the climate patterns of the past to be revealed with amazing precision. These climatologists found themselves studying tree rings, pollen spores, ice cores, ocean sediment cores, ancient hydrological systems, archaeological evidence, wine records, old diaries, and so on. These proxies contained signals about global temperature and precipitation trends over a variety of time scales—we suddenly knew a great deal about the long-term climate history of the planet. Some of these climatologists spoke at length about the nature of long-term climate fluctuations, and in the midst of the cold winters of the 1970s, their conclusions shocked the world.

Figure 1 presents the planetary temperature for the past 850,000 years as revealed from the analyses of seafloor sediment that has built up over this long time period. As can be seen in the plot, planetary temperatures of the past 850,000 years were typically far below those of the most recent few centuries. When compared to the past million years, we presently live in a very warm period that must be considered anomalous given the observed pattern in global temperatures. For another perspective, one should realize that the temperatures of the past 1 million years appear to be far lower than the temperatures for the preceding 40 million years (see Ruddiman and Kutzbach, 1991); the planetary temperatures during the age of the dinosaurs before that were much higher than any over the past million years.

Climatologists in the late 1970s were fully aware that glacial periods dominated the climatic record over the past 850,000 years. However, they noticed that the pattern was interrupted by a few interspersed, short-lived, unusually warm interglacials. These interglacials rarely lasted more than 10,000 to 12,000 years, and the present interglacial

Figure 1.   Reconstructed global temperatures over the past 850,000 years (Shackleton and Opdyke, 1973).

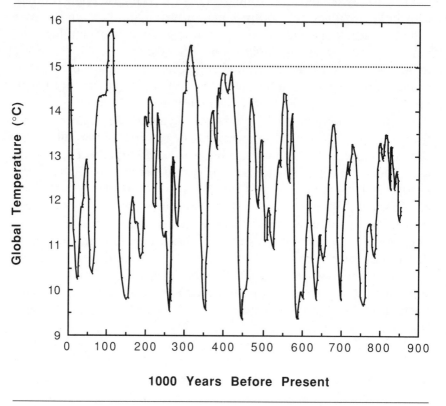

period was approaching 11,000 years old. Some climatologists of the 1970s wondered out loud if we were headed into the next ice age. The planetary temperature appeared to be falling, the Arctic temperatures were dropping quickly, and the United States was experiencing brutally cold winters in the mid to late 1970s (e.g., Diaz and Quayle, 1978, 1980a). Climate changes of the past appeared to be rapid, with enormous declines in temperature over remarkably short time periods. Could the winters of the late 1970s be the signal that we were returning to yet another ice age? According to many outspoken climate scientists in the late 1970s, the answer was absolutely yes—and we needed action now to cope with the coming changes.

Climatology was suddenly front-page news. Leading magazines carried the story on their covers, the message of global cooling was selling newspapers, professional journals in climatology were born, universities and research institutions were hiring climatologists, and funding agencies supported climate projects at unprecedented levels. Observed cooling trends prompted the publication of a number of important and popular books including *The Cooling* (Ponte, 1976), *Forecasts, Famines and Freezes* (Gribbin, 1976), *The Genesis Strategy* (Schneider, 1976), *Climates of Hunger* (Bryson and Murray, 1977), and *The Climate Mandate* (Roberts and Lansford, 1979). *National Geographic* published a full-length article entitled "What's Happening to Our Climate?" in which the pros and cons of global warming or cooling were beautifully presented (Matthews, 1976). The cold winters in the United States and Europe in the mid to late 1970s continued to fuel the growing consensus that we were headed to a much cooler climatic future.

Climatology was alive and well in the 1970s, thanks in no small part to the perceived threat of global cooling. The many problems associated with the cooling were further exacerbated by the compounding linkages to the ongoing "oil crisis." Not surprisingly, we heard the call for action from scientists and decisionmakers—act now on global cooling and avoid the dire consequences of the coming disaster!

However, some scientists were skeptical, and they pointed to a future of global *warming,* not cooling, resulting from a continued buildup of greenhouse gases. These scientists were in the minority at the time, and the media saw little need to cover much of "the other side" of the climate change story. But for more than a century, researchers had written about the direct connection between greenhouse gases and the planetary temperature. If the greenhouse gases are increased, the appropriate and often very simple calculations showed that the planetary temperature should also increase. And by the late 1970s, many measurements were showing an alarming, exponential increase in the concentration of greenhouse gases in the atmosphere. These scientists who were skeptical of the coming-ice-age arguments were well aware of the physical connection between the concentration of these greenhouse gases and the temperature of the planet.

Their argument was relatively simple, and it was well grounded in the physics of the climate system. In the absence of these greenhouse gases, the planetary temperature should reflect a balance between incoming energy absorbed by the earth and the amount of energy emitted away from the earth. The overall calculation comes from the following relatively simple equation:

$$(\pi r^2)S(1-\alpha)=(4\pi r^2)\epsilon\sigma T^4 \tag{1}$$

Starting from the left side, π is the ratio of the circumference of a circle to its diameter, r is the radius of the earth, S is the solar constant—the amount of energy arriving from the sun to the earth expressed as energy per unit area—and α is the albedo or reflectivity of the earth (varying from near 1 for highly reflective white surfaces to near 0 for dark, absorptive surfaces). On the right side, ϵ is the emissivity of the earth (describing the earth's efficiency as a radiating body compared to a perfect radiator or blackbody); σ is the Stefan-Boltzmann constant, which specifies the energy radiated from a blackbody at a given temperature; and T is the planetary temperature. The πr^2 term determines the area of a disk, defined by the size of the earth, that is intercepting the energy from the sun. When multiplied by the solar constant, S, the amount of solar energy intercepted by the earth is determined. When this quantity is multiplied by $(1-\alpha)$, the amount of solar energy *absorbed* by the earth is established.

While the left side of Equation 1 shows the amount of solar energy absorbed by the earth (in the absence of our atmosphere containing greenhouse gases), the right side shows the amount of energy emitted by the earth back to space as back-radiation. The term $\epsilon\sigma T^4$ determines how much energy is emitted per unit area by a body at temperature T. When multiplied by $4\pi r^2$, the area of the earth, we get the total energy emitted by the entire earth. Because a body in space tends to maintain a radiation balance, we can solve the equation for T and get:

$$T = \sqrt[\cong]{\frac{(\pi r^2)S(1-\alpha)}{(4\pi r^2)\epsilon\sigma}} = \sqrt[\cong]{\frac{S(1-\alpha)}{4\epsilon\sigma}} \tag{2}$$

A planetary temperature of approximately $-20°C$ ($-4°F$) is calculated when the appropriate values are inserted and the equation is solved for T. Yet, the global average temperature is much higher; the best estimate of the actual temperature of the earth is near $15°C$ ($59°F$). The reason for the difference is that the various gas molecules in the atmosphere absorb energy of certain wavelengths and re-emit it in all directions. The greenhouse gases absorb and re-emit the infrared energy of the earth's back-radiation, delaying its return to space. Some of this energy is re-emitted straight upward into space, some goes downward and is reabsorbed by the earth, and some is absorbed by other gases and particles in the atmosphere. All of it eventually returns to space, leaving the radiation balance intact, but the numbers on the right side of Equation 1 change—the net result is that the temperature T increases. The total effect is like adding a layer of insulation to the earth.

For a given reflectivity of the planet (assumed for the sake of these calculations not to vary with or without greenhouse gases), these equations show that a greenhouse effect produces a 30°–35°C (54°–63°F) increase in the global temperature. This naturally occurring greenhouse effect has existed throughout the evolution of Earth's atmosphere, the same effect appears on other planets of the solar system (Venus has its "runaway" greenhouse effect), and the underlying principles of the greenhouse effect have been reasonably well understood for centuries.

Oddly, the "greenhouse" effect is actually a misnomer. Glasshouses maintain their higher temperatures by not allowing heated air within the enclosure to escape through convective mixing. Glasshouses are not heated by trapping radiant energy (as in the free atmosphere), but rather, by preventing the normal convective processes (see Lee, 1973; Arnfield, 1987). Nonetheless, the "greenhouse" label has been around for decades, and despite its technical inaccuracy, the label is likely to be around for years to come.

The scientists of the late 1970s were fully aware of these greenhouse principles, but the planet appeared to be cooling either from natural fluctuations in the climate system or from some external forcing mechanism. The observed cooling may have resulted from some small change in the orbital characteristics of the earth or possibly from the release of particulates into the atmosphere by human activities. Phrases like "the human volcano" were in vogue in the environmental community at that time. But despite the cooling trends reported for the globe over the thirty or so years prior to the late 1970s, some scientists warned that the continued buildup of the greenhouse gases may one day cause significant warming of the planet (e.g., Manabe and Wetherald, 1975; Schneider, 1975; Wang et al., 1976; Ramanathan et al., 1979). As we moved into the 1980s, the jury was still out, but those scientists predicting continued cooling appeared to have an upper hand.

The 1980s saw phenomenal growth in the field of climatology. The great excitement of the mid to late 1970s had been translated into a vibrant field of scientists who proudly called themselves climatologists. Advances in computer hardware and software allowed for enormous strides in (1) computer modeling of the climate system, (2) storage and utilization of giant climate data bases, and (3) multivariate statistical analyses of the variations found in the climate records. Satellite systems were moving into more advanced, operational stages, and the climatologists produced large numbers of research articles in a growing family of professional journals. The 1980s also saw the end to any global cooling trends reported earlier; record-breaking warmth throughout the globe began to make the headlines. Momentum for the "popular vision" of the greenhouse effect was growing—the improved climate models were predicting warmer temperatures with increased concentrations of greenhouse gases, and the planetary temperature appeared to be on the rise. The issue exploded, particularly in the United States, in the summer of 1988.

The Summer of 1988

Before the summer of 1988, the scientific journals contained greenhouse-related articles that generally fell into two broad categories. One category was articles reporting the results of numerical modeling experiments suggesting that a doubling of carbon dioxide would raise the planetary temperature 4°C (7°F) or more (e.g., Manabe and Wetherald, 1980; Hansen et al., 1981; Wetherald and Manabe, 1981, 1986; Mitchell, 1983; Washington and Meehl, 1983, 1984; Wigley and Schlesinger, 1985; Wilson and Mitchell, 1987). The other category of articles grew out of the development of records of hemispheric and global temperature extending for more than 100 years (e.g., Jones et al., 1982, 1986a, 1986b, 1986c; Hansen and Lebedeff, 1987). Analysis of these temperature data suggested that the world was warming at a rate of about 0.5°C (0.9°F) per century. The 30-year cooling that had been noted by the scientists of the late 1970s was imbedded in a general upward trend, and that 30-year cooling was largely restricted to the Northern Hemisphere. The greenhouse effect was alive in the scientific journals, but not yet alive in the popular media.

By 1988, the stage was set to unleash the greenhouse effect on the public and the decisionmakers. The summer of 1988 was dominated by extremely high temperatures and drought throughout much of the United States, and the climate community once again found itself in the media spotlight. On June 23, 1988, at a United States Senate hearing on climate change, James Hansen (director of NASA's Goddard Institute for Space Studies) reported that the world was warmer than at any time in the instrumental record and that some part of the recent warming was likely a result of the buildup of greenhouse gases. He spoke of being "99 percent certain" that some of the observed changes in climate were linked to the greenhouse effect. Hansen became a recognizable face; he was seen as the man who knew why we were so hot and dry, who warned us of the impending climate disaster that could result from our activities. In the midst of the summer heat and drought, the momentum for the greenhouse issue increased, and an endless number of scientists, politicians, and decisionmakers eagerly adopted the increasingly distorted message. Predictions of climate-related disasters filled the newspapers, magazines, and airwaves; climatology was on our minds again.

If the volume of scientific evidence was not sufficiently convincing, the summer heat of 1988 was likely to win over any remaining greenhouse skeptics. And if the summer heat was still insufficient, two other events of 1988 seemed to provide the nail for the skeptics' coffin. In July of that year, a few wildfires were burning in Yellowstone National Park that looked quite normal given the long-term fire record for the area. But by July 21, the wildfires were expanding through the parched forests of the park, and managers decided to suppress all further ignitions for the summer. On August 20, a cold front whipped through the Yellowstone area with high winds and lightning, but little rainfall. The serious situation perceived by park managers had become a calamity—the wildfires of Yellowstone appeared to be out of control. Almost 400,000 hectares (nearly 1,000,000 acres) of the park burned, a value almost *50 times larger* than the burn area for any year during the period 1895–1988 (Christensen et al., 1989; Romme and Despain, 1989). Yellowstone Park wildfires were front-page news, and the link to the greenhouse was a natural. The greenhouse effect would raise the temperature, increase the aridity, weaken the forest, and promote an increase in wildfire activity. The Yellowstone Park fire, along with many other fires in the American West during the summer of 1988, became an integral part of the "popular vision."

As the fires of the Yellowstone area were dying in September (some continued into November), yet another event occurred that appeared to be consistent with the greenhouse predictions. Some scientists (e.g., Emanuel, 1987a, 1987b) had suggested that global warming would increase tropical sea-surface temperatures, and ultimately, these higher sea-surface temperatures would support larger storms in the tropical atmosphere. Emanuel (1987a, 1987b, 1988) even spoke of enormous hurricanes called "hypercanes" that could arise in a greenhouse world.

On September 9, 1988, Tropical Storm Gilbert was growing south of the Virgin Islands. As the storm trekked west-northwest, its pressure continued to drop, and its sustained winds increased at an alarming rate. As fully developed Hurricane Gilbert approached the Yucatán its pressure had fallen to 885 millibars in the eye, the lowest pressure ever recorded for a Western Hemisphere hurricane. With maximum sustained winds near 280 km/hr (175 mph), Gilbert smashed into the Yucatán, including the highly developed recreational area around Cancún. Once again in the summer of 1988, the weather was leading the news. By September 16, Gilbert struck northeast Mexico and southern Texas with high winds and torrential rains (Eden, 1988). It was downgraded to a tropical storm the next day, but the enormous damage the storm created appeared, in the eyes of many, to be linked directly to the greenhouse effect.

Thanks to this relatively brief period of atmospheric extremes, everyone knew about the greenhouse effect, and everything that followed, from large windstorms in London to low water levels on the Mississippi River to earthquakes in California, was somehow linked to the greenhouse issue. The argument was relatively simple: the developed world was pumping greenhouse gases into the atmosphere (and deforestation was reducing the uptake of carbon dioxide), the atmosphere was warming, the changes in atmospheric circulation were causing unprecedented events to occur in the climate experienced at the surface, and we must act immediately to reduce the emission of greenhouse gases or face the calamities of 1988 over and over again. We had to save the earth, and the greenhouse effect became the center-piece in the call for action. Award-winning scientists believed in the greenhouse effect, the top scientific journals were publishing landmark articles supporting the "popular vision" (e.g., Hansen et al., 1988; Jones, 1988; Kellogg and Zhao, 1988), the arguments in favor of global warming had a solid theoretical base, and the climate system appeared to be producing clear greenhouse signals. However, by the end of 1988, many scientists were becoming increasingly skeptical about the apocalyptic view of the greenhouse effect; these "skeptics" were begin-ning to steal some of the spotlight.

THE EMERGING SPECTRUM: EDEN OR APOCALYPSE?

Inevitably, an issue like the greenhouse effect quickly gains enormous support at first. And initially, the message of the greenhouse effect was conveyed with a sense of overwhelming support in the scientific community. As the issue grew in perceived importance, a number of reporters began to search for the other side of the issue. And to their amazement, the same scientific journals with articles supporting the greenhouse effect contained articles seriously challenging the foundations of the apocalypse. Just as many good scientists supported the popular view of global warming, a number of reputable scientists questioned the very underpinnings of the emerging greenhouse paradigm.

By the end of 1988, an interesting but predictable spectrum had developed for the greenhouse issue. At one end of the spectrum was an army of scientists, politicians, decisionmakers, movie stars, rock stars, and environmentalists who embraced the apocalyptic view of the greenhouse effect. In too many circles, the greenhouse effect took on a religious quality; believers in the issue seemed entirely insensitive to any suggestion that the greenhouse paradigm was in error. They argued that irrespective of any developing scientific debate, the chance of catastrophe was real, policies had to be adopted to reduce the emission of greenhouse gases, deforestation had to be reversed—we could not afford *not* to believe in the greenhouse effect. Even if they were wrong about the climate response to increasing greenhouse gases, they believed that their suggested policies were so good for the planet that the policies should be adopted irrespective of the outcome of the greenhouse effect.

At the other end of the spectrum were some scientists, a few politicians, a few decisionmakers, and absolutely no movie stars, rock stars, or environmentalists. These people argued that the climate changes resulting from the buildup of greenhouse gases would likely be very small, and some even beneficial. They argued that decreasing afternoon temperatures and increasing nighttime temperatures would reduce thermal stresses in the biosphere, the increased cloudiness and rainfall would increase soil moisture and reduce the threat of drought, icecaps would not melt, severe storms may *decrease* in intensity and frequency, and the climate system would be less inclined to produce large fluctuations. The increase in carbon dioxide could aid many plants in their growth, more crops could be produced and yields would increase, and less water may be needed for their development. After

4 billion years of evolution, nature had not made a fatal error, and mankind was not the curse of the earth. This group also tended to believe that most policy options proposed to deal with the greenhouse threat were naive and not likely to have much impact on future concentrations of the various greenhouse gases.

Both extremes of the spectrum appear to find solid scientific evidence to support their views. Not surprisingly, a majority of scientists fall somewhere between these two extremes. These scientists attempt to recognize the validity of arguments within the spectrum of greenhouse research, they attempt to identify causes of such disparate conclusions, and they live in a research environment of constant questioning, debate, and re-evaluation. These people rarely find themselves in the spotlight, but they are the ones who will resolve the critical questions, and they are the scientists who, ultimately, will shape the way humanity responds to the greenhouse threat. These scientists will uncover the inescapable facts of the greenhouse effect. And while in the short run the players at the extremes of the spectrum will appear to dominate the debate, the many careful and creative scientists in the field who have not adopted (and then carefully guarded) an extreme position will emerge as the true impact players.

THE SEARCH FOR ANSWERS
AND "PROBABLE" CLIMATE CHANGE

The purpose of this book is not to win readers over to one side or another in the greenhouse debate. Rather, its purpose is to explore the range of outcomes that may result from the buildup of greenhouse gases. Recognizing that we now have approximately 100 years of climate records from throughout the globe, and recognizing that the past 100 years have witnessed a tremendous buildup of greenhouse gases, we will explore what changes in the climate system have occurred during this historical period.

Some of these changes in climate are very consistent, others completely inconsistent, with the predictions of greenhouse theory. And the predictions themselves are extremely varied, providing an enormous range of greenhouse possibilities. The highest confidence is placed on the predictions that are consistent with the observations, and the lowest confidence is placed on the predicted outcomes that seem completely inconsistent with our observations of the climate system. A view of future climate change is developed that is consistent with both the theoretical predictions from the greenhouse research and empirical climate observations made throughout the world.

At times this composite view of climate change will appear to be very consistent with the mainstream on the issue, while at other times the view will be closer to the extremes. The book is an invitation to (1) drop any rigid attachment to one side of the debate or the other, (2) begin to explore the complexities of the debate, (3) learn to appreciate the strengths and weaknesses of the various arguments, and (4) place more thought in the climate component of policies that deal with global change. In too many issues, people who have read the least have the greatest confidence in their convictions. The more we learn about an issue, the more we appreciate the various sides of a debate. Initially, our confidence in the predicted outcome deteriorates. After even more reading and exploring, one side or another will become more consistent with *your* understanding of the issue, and your confidence and conviction will rise. This book is written to move you closer to your own conclusions about the climatic consequences of the greenhouse effect.

After reading the book, you still will not be certain of the climate outcome of increasing the greenhouse gases. Uncertainty will always be a major part of climate predictions for the future; however, the evidence presented here may allow some of the uncertainty to be reduced. Our decisions to plant a tree, impose a carbon tax on fossil fuels, provide nuclear power to some developing nation, or entirely restructure our social and economic foundations in order to cope with the greenhouse effect must be based on the best science available. And the best science, unfortunately, will be loaded with uncertainties. We all must decide if we want our greenhouse policies to be based on greenhouse hype or on the best science available. Herein lies an inescapable dilemma—decisionmakers may be forced to build greenhouse-related policies in an environment of substantial scientific uncertainty. I hope this book will help you understand and even narrow some of the uncertainty surrounding the heated greenhouse debate.

2

THE EXPONENTIAL INCREASE IN GREENHOUSE GASES

The greenhouse effect has been a natural part of the climate system for billions of years. The earth produced copious quantities of carbon dioxide and other greenhouse gases from the earliest moments in its evolution. And as we shall see, the planet itself has the ability to generate large fluctuations in the concentrations of these gases, without any interference by human activities. In terms of billion-year time scales, some greenhouse gases are presently near their *lowest* concentrations while others are probably at the highest concentrations ever seen on the earth. Understanding the types, origins, concentrations, and rates of increase in these greenhouse gases is critical to developing an appreciation of the climate consequence of their continued buildup into the next century.

This chapter will deal with the greenhouse gases that are added to the atmosphere by human activities; these are often referred to as the anthropogenic greenhouse gases. It should be noted from the outset that water vapor is by far the most significant of the greenhouse gases. If a total greenhouse effect is defined as the sum of the influence of water vapor and all other greenhouse gases, then water vapor accounts for nearly 70 percent of the overall radiative effect (Shine et al., 1990). Without water vapor in the atmosphere, along with the water droplets in clouds, the overall greenhouse effect would be much smaller than what is observed. The natural fluxes of water vapor exchanges between the oceans, land areas, and atmosphere are astronomical in size, and the direct influence by humans is thought to be small on a global scale. However, the atmospheric concentration of the minor greenhouse gases is substantially affected by human activities; the changes in concentrations of carbon dioxide, methane, nitrous oxide, the chlorofluorocarbons, and other gases are thought, by the greenhouse scientists, to lead to the climatic changes associated with the "popular vision."

So in essence there are two meanings of the "greenhouse effect." On one hand, the earth's temperature is maintained by the trapping of infrared radiation by water vapor and the remaining minor trace gases. This impact on global temperature is referred to in the atmospheric sciences as the "greenhouse effect," and climatologists agree that the earth's temperature is maintained by this process. However, today the "greenhouse effect" has come to be defined as the change in climate that will result from human activities pumping trace gases into the atmosphere. Throughout the rest of the book, unless otherwise clearly stated, "greenhouse effect" describes human-induced changes in climate brought about through the increase in the various greenhouse gases.

CARBON DIOXIDE

Carbon dioxide (CO_2) and the greenhouse effect are terms used almost interchangeably by scientists and others who are interested in global change issues. Indeed, many believe that the projected climate changes related to the greenhouse effect stem solely from the buildup of CO_2 caused by our burning of fossil fuels and from our destruction of the tropical rainforests. As we shall learn, CO_2 has shown enormous natural variations over long time scales, its atmospheric concentration is increasing exponentially at the present time, and CO_2 contributes only about half of the anthropogenic greenhouse effect.

Long before the evolution of humans, enormous quantities of CO_2 were pumped into the atmosphere from erupting volcanoes and the outgassing from geysers along thermally active geological zones. The early atmosphere of the earth must have been highly enriched in CO_2; the concentration was likely to have been tens to hundreds of times— one or two orders of magnitude—higher than the modern-day concentration (Hart, 1978; Holland, 1984). With continued eruptions and outgassing, the earth must have appeared to be heading for the life-prohibiting runaway greenhouse effect. Luckily, the planetary system had a number of ways for CO_2 to be withdrawn from the atmosphere thereby stabilizing and then reducing the atmospheric CO_2 levels.

At the same time CO_2 was pumped into the atmosphere, large amounts of water vapor also entered the system. The vapor condensed into clouds, and inevitably, these primeval clouds produced precipitation that began to fill the world's ocean basins. Because CO_2 in the atmosphere dissolves in rainwater as carbonic acid, significant quantities of CO_2 were extracted from the atmosphere into the developing oceans. There it combined with other ions in the water to form carbonates, which when conditions were right came out of solution into solid form and settled to the ocean floor as inorganic limestone. The oceans and carbonate rocks became large reservoirs of carbon. Much later in geologic history, living organisms evolved ways of building shells and skeletons from carbonates. As these organisms died and their skeletons rained to the seafloor, great beds of organic limestone were piled up. These limestone beds, and later still coal beds—essentially pure carbon from the remains of land plants— sequestered huge quantities of carbon and helped keep atmospheric CO_2 levels low.

The continued development of life saw the establishment of a balance among CO_2 exchanges in the planetary system. The details of the earth's carbon dioxide balance are provided in a number of sources (e.g., Trabalka, 1985; Post et al., 1990; Watson et al., 1990; Jenkinson et al., 1991), but the basic principles are relatively simple. The atmosphere gains its CO_2, in decreasing order of importance, from (1) a surface exchange of CO_2 with seawater governed by wind speed, temperature of the air and seawater, and various gas transfer coefficients (Tans et al., 1989), (2) animal and plant respiration (this includes the CO_2 produced by exhaling by animals and the decay of plant materials), (3) combustion of fossil fuels, (4) direct release of CO_2 from the soil, and (5) the outgassing from volcanoes, hot springs, and related geological features. The atmosphere loses CO_2, in decreasing order of importance, (1) by the surface exchange with the oceans (some of this carbon is further removed by phytoplankton within the sea), (2) by photosynthesis, and (3) to a much lesser degree, by the weathering of rocks and the development of new fossil beds. The oceans remain by far the largest reservoir of CO_2 followed by geological formations, the biosphere, and the atmosphere.

Without significant interference by humans, one may expect the carbon cycle to achieve a balance in which the concentrations and exchange rates of CO_2 would stay about the same over long periods of time. However, natural variations in volcanism, climate, vegetation distributions, and oceanic circulations create large variations in atmospheric CO_2, irrespective of human activities. In order to capture a glimpse of CO_2 levels from long ago, we are fortunate that bubbles of air were trapped in the ice collecting in Greenland and Antarctica, among other places. When deep ice cores are analyzed, the air from thousands of years ago reveals preexisting CO_2 concentrations. One very famous analysis by Barnola et al. (1987, 1991) from the ice drilled near the Vostok station in Antarctica shows that over the past 160,000 years, atmospheric concentration of CO_2 ranged from less than 180 parts per million (ppm) by volume 42,000 years ago to almost 300 ppm 134,000 years ago (Figure 2). The plot of reconstructed CO_2 levels shows tremendous variability over the 160,000 years, including a noticeable trend upward beginning about 16,000 years ago.

Figure 2.   Atmospheric concentration of CO_2 over the past 160,000 years from the Vostok ice core (from Barnola et al., 1987).

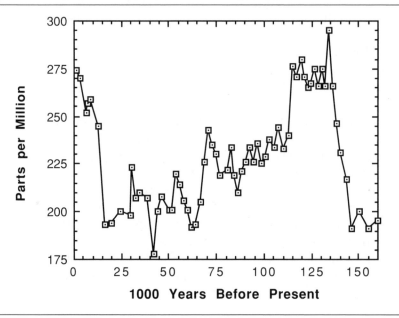

The Industrial Revolution substantially changed the carbon dioxide cycle. Prior to the 1750s, the amount of CO_2 appeared to be near the 275–280 ppm level (the Vostok record shows CO_2 concentrations to be 270 ppm about 3,350 years ago and 274.5 ppm approximately 1,700 years ago). The Industrial Revolution spawned not only a rise in our combustion of fossil fuels, thereby adding to the atmospheric CO_2 level, but also a significant rise in the human population. We began increasing our CO_2 emission per capita at the same time our numbers were growing; the result was an enormous increase in CO_2 emission. A plot of global CO_2 emission rates from Rotty and Masters (1985) clearly shows a near linear increase in emissions of CO_2 from 1880 to about 1950 (Figure 3). Following 1950, the rate of increase is still rather linear, but almost seven times faster than the rate previous to

Figure 3.   Global carbon dioxide emission rates (units are 10^{15} grams of carbon per year), 1880–1988 (from Rotty and Masters, 1985, and Boden et al., 1990).

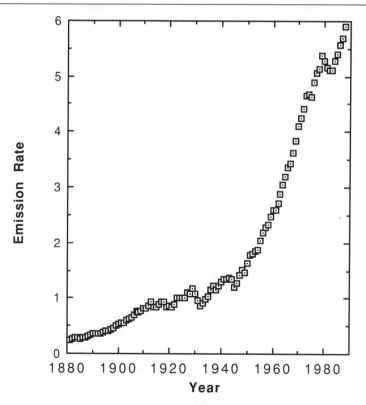

1950; the pattern is noticeably interrupted by the oil crisis of the mid-1970s. Although these two segments defined by a break around 1950 are somewhat linear, the overall pattern for the past 100 years is best described as exponential.

Due to this human injection of CO_2 into the system, the concentration of CO_2 in the atmosphere began to rise. However, the rise was less than would have been expected if a one-to-one relationship existed between CO_2 released by humans and the CO_2 increase in the atmosphere. The storage capacity of the ocean and biosphere is able to absorb some of this increase; nevertheless, the human injection has overwhelmed the absorption capacity of the system, and atmospheric CO_2 is on the rise. Figure 4 shows the rise in atmospheric CO_2 over

Figure 4.   Atmospheric carbon dioxide concentrations from 1764 to 1988. The open boxes are measurements made from ice recovered at Siple Station, Antarctica (Raynaud and Barnola, 1985), and the closed boxes are annual CO_2 concentrations measured at Mauna Loa, Hawaii (Bacastow et al., 1985). Actual data are available from Boden et al. (1990).

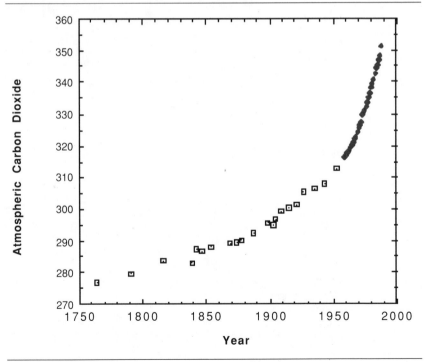

the past few centuries as determined from analysis of an ice core taken at Siple Station, Antarctica, and direct atmospheric measurements made more recently at Mauna Loa Observatory in Hawaii (Raynaud and Barnola, 1985; Bacastow et al., 1985). The plot shows that CO_2 levels have risen 20 percent over the past 100 years (from 293 ppm in 1888 to 351 ppm in 1988) and 27 percent since the beginning of the Industrial Revolution (from 277 ppm to 351 ppm). Given the exponential nature of the curve, the rate of increase is itself increasing at a startling rate. And these CO_2 molecules in the atmosphere do not go away easily— the average CO_2 molecule has an atmospheric lifetime of between 50 and 200 years (Watson et al., 1990). There is no debate here—the atmospheric concentration of CO_2 is rising fast, and the anthropogenic emissions are causing the increase.

Inevitably the question arises, just who is responsible for this gigantic increase in atmospheric CO_2? The industrialized nations have been the major contributors to date, but the developing countries are assuming a larger role as their populations and demands for energy increase. Figure 5 shows that approximately half of the 1988 emissions of CO_2 from fossil fuels came from the United States, USSR, and China. Given development plans of the various nations, China, India, and the Soviet Union are almost certain to increase their fossil fuel emissions of CO_2 in the immediate future.

Other charts reveal interesting geopolitical patterns in our emission of CO_2 from fossil fuel consumption. Figure 6 shows the 1988 CO_2 emission rate per person for a variety of different countries. The figure shows that the global average of 1.2 metric tons of CO_2 per person is exceeded by almost every country in these plots; obviously, many countries not shown in Figure 6 are presently well below the global average. Particularly noticeable is the fact that in 1988, East Germany, the United States, and Canada were all approximately four times higher than the global average. This inescapable fact is repeatedly used in the call for immediate action. As we saw earlier, the "popular vision" sees a catastrophe about to occur, and now we see who is responsible for the dangerous increase in CO_2. According to many followers of the "greenhouse gospel," these nations with the highest per capita CO_2 emissions must act now to lower their contribution to the greenhouse effect.

However, a very different pattern emerges when the CO_2 emission rates are related to the Gross National Product (GNP) of these nations

Figure 5.   Percent of 1988 global CO$_2$ emission from fossil fuels by selected countries (from Boden et al., 1990).

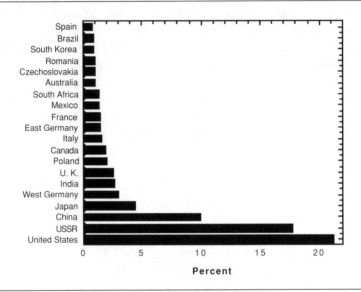

Figure 6.   1988 CO$_2$ emission rate per person by selected countries (from Boden et al., 1990).

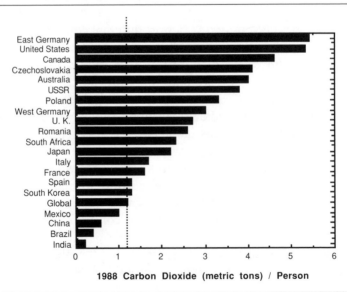

(Figure 7). The 1988 global per capita GNP is estimated to be $3,470 (Haub et al., 1990); the global emission of CO_2 per capita is near 1.2 metric tons (Boden et al., 1990). Therefore, the 1988 global average CO_2 emission per GNP is approximately 0.35 metric tons of CO_2 release for every $1,000 GNP. The United States and Canada, with their very high rates of CO_2 emission per person, actually are *below* this global average. The United States may produce an enormous amount of CO_2, but it also produces a very high GNP. While the globe as a whole releases 0.35 metric ton of CO_2 to produce $1,000 of GNP, the United States releases only 0.27 metric ton to produce the same $1,000 of GNP. The global average in this measure is nearly 30 percent higher than the value obtained for the United States. It may seem counterintuitive or just unacceptable, but the United States is much more efficient in generating GNP at lower CO_2 emissions than most countries. One way to reduce CO_2 emissions substantially is to make much of the world as "efficient" as the United States. Although there may be any number of arguments against using GNP versus CO_2 as a measure of national efficiency, these simple charts nonetheless show interesting patterns in the emission of CO_2.

Figure 7.   1988 CO_2 emission rate per Gross National Product ($1,000) by selected countries (emission data are from Boden et al., 1990, and GNP data are from Haub et al., 1990).

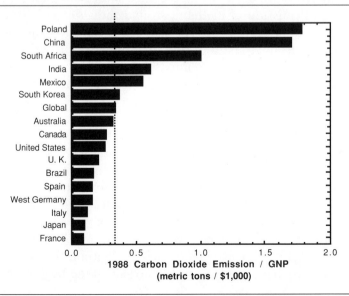

There is no doubt about it—human activities are substantially increasing the atmospheric concentration of CO_2, and these activities will certainly continue to force atmospheric CO_2 levels to climb. In the community of greenhouse scientists, there is virtually no debate surrounding this observed increase of atmospheric CO_2. The rate of increase has been measured throughout many parts of the globe, the rate is recognized to be exponential over the past 100 years, and the emission rates by nations are reasonably well known. Given the physical reality that CO_2 in the atmosphere acts to warm the earth by absorbing energy emitted by the earth's surface and atmosphere, one may safely conclude that the increase in atmospheric CO_2 will act to warm the earth to some degree.

METHANE

Although carbon dioxide represents the most important anthropogenic greenhouse gas, many other gases, including methane (CH_4), contribute significantly to the greenhouse effect. Methane enters the atmosphere by a number of sources associated with anaerobic biological processes (those which occur in the absence of oxygen). Major sources of methane include natural wetlands, rice paddies, and the enteric fermentation in the stomachs of ruminant animals (cattle, sheep, and wild animals). When combined, these three sources account for 58 percent of total emission of methane around the globe (Watson et al., 1990). Other important sources of methane include biomass burning, termites, gas drilling, landfills, and coal mines (Crutzen, 1991).

Most of the methane emitted into the atmosphere undergoes a complex reaction with carbon monoxide, nonmethane hydrocarbons, nitrogen oxide, ozone, and hydroxyl ion (Sze, 1977). As a result methane has an atmospheric lifetime of about a decade, which is 5 to 20 times less than the lifetime for carbon dioxide. Nonetheless, the atmosphere has been unable to eliminate the large increases in methane, and resultant concentrations are rising quickly.

Methane concentrations, like the concentrations of CO_2, have increased exponentially over the past few centuries (Figure 8). Methane concentrations were near 0.75 ppm in 1800; however, the recent measurements show methane levels to be near 1.70 ppm (Stauffer et al., 1985; Blake and Rowland, 1988). Concentrations of atmospheric methane have more than doubled since the beginning of the

**Figure 8.   Atmospheric methane concentrations (ppm) from 1771 to 1988.
The open boxes are measurements made from ice recovered at Siple Station,
Antarctica (Stauffer et al., 1985), and the closed boxes are annual CH_4
concentrations measured at a variety of stations around the globe (Blake
and Rowland, 1988; Khalil and Rasmussen, 1990). Actual data are available
from Boden et al. (1990).**

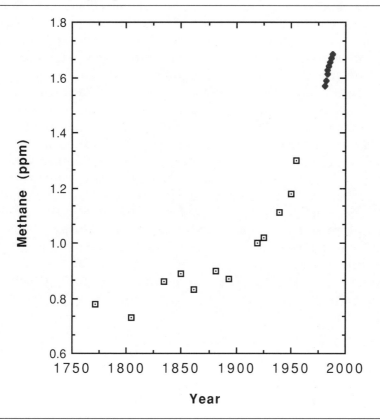

Industrial Revolution, and the levels are likely to continue to rise at
an accelerated pace.

Unfortunately, the data do not exist for establishing the emission
rates of methane by individual countries. However, given the vastly
different sources of methane when compared to sources of CO_2, we
could safely conclude that the patterns discussed for CO_2 would be
substantially different for methane. Countries like China and Vietnam,

with their extensive development of rice paddy agriculture, would likely emerge as major contributors of methane to the atmosphere.

NITROUS OXIDE

Nitrous oxide (N_2O) is yet another naturally occurring greenhouse gas that has increased in atmospheric concentration due to human activities. Unlike carbon dioxide and methane, the sources and sinks of nitrous oxide are not well understood, nor are the magnitudes of the various exchanges. The oceans appear to be a large source of N_2O (Watson et al., 1990); however, significant emission of N_2O occurs from the soils in tropical forests (Keller et al., 1986), and in particular, from soils that have undergone deforestation. Soils in mid-latitude forests, combustion of fossil fuels, biomass burning, and the use of nitrate and ammonia fertilizers also supply N_2O to the atmosphere. Nitrous oxide that enters the atmosphere is either taken up by the soil or ultimately destroyed by photochemical decomposition in the stratosphere; however, the N_2O molecules have atmospheric lifetimes near 150 years. As with CO_2 and CH_4, the ability of the earth-atmosphere system to absorb N_2O has been overwhelmed by the recent emission rates. As a result, atmospheric concentrations of N_2O have risen from about 285 parts per billion (ppb) for pre–Industrial Revolution levels to approximately 310 ppb in 1990 (Khalil and Rasmussen, 1983; Watson et al., 1990).

CHLOROFLUOROCARBONS

Carbon dioxide, methane, and nitrous oxide are all greenhouse gases that have been a natural part of the atmosphere for billions of years. As we have seen, relatively recent human activities have increased the emission of CO_2, CH_4, and N_2O into the atmosphere, the system's absorption capacity has been overwhelmed, and the atmospheric concentrations of these gases have increased. Unlike these other greenhouse gases, the pre–Industrial Revolution atmospheric concentration of the chlorofluorocarbons (CFCs) and a wide variety of related gases was essentially zero. Virtually every CFC molecule found anywhere in the atmosphere today can be traced back to an anthropogenic emission.

The CFCs originate from aerosol propellants, refrigerants, foam blowing agents, various industrial solvents, and fire retardants. These gases enter the atmosphere, eventually move into the stratosphere, and are finally destroyed by photodissociation. The process of CFC removal takes decades to centuries, and as a result, the rate of increase in atmospheric CFC levels is greater than any other greenhouse gas. These CFCs are very powerful greenhouse gases, and they are not easily removed from the atmosphere. Despite having concentrations that are measured in parts per trillion, the CFCs add significantly to the overall greenhouse effect, more than methane and nitrous oxide combined.

Recognizing a number of potential hazards associated with the CFCs (greenhouse effects, possible depletion of the ozone layer), the Montreal Protocol on Substances that Deplete the Ozone Layer has attempted to produce international agreements to reduce and ultimately phase out the emissions of CFCs and related gases. Nonetheless, given the long residence time of the CFCs and their continued emission from a number of sources, atmospheric concentrations of the CFCs are likely to be significant well into the next century (Watson et al., 1990).

EQUIVALENT CARBON DIOXIDE

Carbon dioxide, methane, nitrous oxide, and the chlorofluorocarbons are all anthropogenically generated greenhouse gases that have been increasing due to human activities. Other greenhouse gases may also be increasing due to anthropogenic activities (e.g., tropospheric ozone, carbon monoxide, NO_x), but their rates of change and impact on the overall greenhouse effect are very difficult to quantify. However, each of the greenhouse gases has a unique and reasonably well known signature in affecting the radiation balance of the earth. Some of the gases, such as the CFCs, are extremely effective on a molecule-by-molecule basis in their ability to trap infrared energy; many of the CFCs are 1,000 to 10,000 times more effective than CO_2 in producing a greenhouse effect. Fortunately, the concentration of the CFCs is relatively low when compared to the concentrations of CO_2. On the molecule-by-molecule basis, methane is over 20 times and nitrous oxide is more than 200 times more powerful than CO_2 in producing a greenhouse effect.

When the concentration of the gases is combined with the radiative properties of the molecules, we can determine the relative importance of each of the major anthropogenic greenhouse gases. CO_2 is by far the most important greenhouse gas, followed by the CFCs, methane, and nitrous oxide (Figure 9). Several other greenhouse gases may

Figure 9.   Relative contribution of various gases to the increased greenhouse effect in the 1980s (Shine et al., 1990).

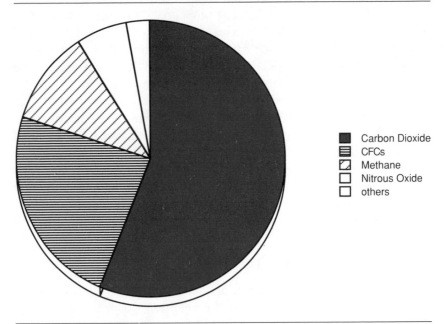

further contribute to the effect, but their contribution is difficult to quantify. It is again worth noting that water vapor is more than twice as important as all other greenhouse gases in producing a greenhouse effect. Should the water vapor increase in the future, we could logically expect an enhancement of the greenhouse effect, in the absence of some related increase in cloud cover. However, unlike water vapor, the gases shown in Figure 9 are unique in that their concentrations are increasing directly due to human activities.

Recognizing the climate impact of these assorted greenhouse gases, a numerical value is needed that summarizes the effect of these gases into one greenhouse indicator. The *equivalent carbon dioxide* value is computed by converting the contribution of each gas into a more generalized carbon dioxide equivalent value. The calculation involves the concentration of the various greenhouse gases along with the radiative properties of the individual molecules (e.g., Wigley, 1987; Schneider, 1989b; Lashof and Ahuja, 1990; Michaels, 1990; Houghton et al., 1990). The calculation is somewhat complex, but the resultant value gives some indication of how much CO_2 would be required to produce the same greenhouse effect as all of the anthropogenic trace gases (CO_2, N_2O, CH_4, CFCs) found in the atmosphere. The resulting equivalent carbon dioxide values are plotted in Figure 10.

Recently, Wang et al. (1991) challenged the use of equivalent CO_2 to represent the combined effects of all trace gases. They rather fairly argued that each of the trace gases has a unique signature in its effect on the radiation balance of the atmosphere. Each of the greenhouse gases uniquely influences infrared radiation of different wavelengths and each operates most efficiently at different levels of the atmosphere.

Figure 10.   Actual CO_2 measurements from the Siple Station (boxes) and Mauna Loa (plus signs) data (see Figure 4) and equivalent CO_2 levels (open circles). Equivalent CO_2 levels are available in a number of sources including the "Policymakers Summary" of Houghton et al. (1990).

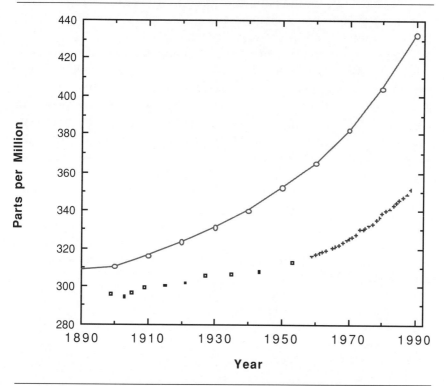

When the many greenhouse gases are introduced into the free atmosphere, they begin to combine with one another, thereby producing another set of complicated interactions with the radiation balance of the system. Scientists recognize that a full representation of each gas is needed to properly account for the radiative effects of the many trace gases. But in an attempt to simplify this complex situation, the equivalent CO_2 values remain in wide use by climatologists working with the greenhouse effect.

Equivalent CO_2 levels were approximately 290 ppm at the beginning of the Industrial Revolution; by 1900, the equivalent CO_2 had risen to about 310 ppm. Although the estimates in the scientific literature vary (e.g., Tricot and Berger, 1987; Schönwiese and Runge, 1991), the best estimate of equivalent CO_2 for 1990 is over 430 ppm—since the beginning of the Industrial Revolution, we have increased the equivalent CO_2 by approximately 50 percent (Houghton et al., 1990). And over the past 100 years, we have seen the equivalent CO_2 levels increase by 40 percent. Given the pattern of the past 100 years, we can expect to reach the 600 ppm equivalent CO_2 value (often used as the value for a doubling of CO_2) between 2035 and 2040.

The concept of equivalent CO_2 is a critical component to much of this book. Most of the predictions associated with the "popular vision" are for a time when we have doubled CO_2 (or equivalent CO_2). Yet, as can be seen in Figure 10, we have already gone halfway to an equivalent CO_2 doubling, and in the past 100 years, we have witnessed a 40 percent increase in this value. We are lucky that over the same 100-year period, relatively good records have been kept regarding the climate of the earth. If large, catastrophic changes in climate are going to occur for a doubling of equivalent CO_2, we should expect to see some of these changes being revealed for a 40–50 percent increase in the equivalent CO_2. Understanding how our climate responded to the observed increase in equivalent CO_2 will certainly provide insight into how the climate will ultimately respond to a doubling of CO_2.

3

THE NUMERICAL MODELS
OF GLOBAL CLIMATE

It is widely recognized that the atmospheric concentrations of the various anthropogenic greenhouse gases are increasing and that they will continue to increase into the next century. As we saw earlier, the doubling of equivalent CO_2 (actually, when we reach 600 ppm) will likely occur near the year 2040; obviously, any number of social, technological, political, and economic unknowns can alter the exact time. However, most scientists agree that some time in the middle of the next century, the earth's atmosphere will reach the 600 ppm level for equivalent carbon dioxide.

For a long time, climatologists have attempted to determine what the climate of the earth would be like in a world of doubled CO_2. Toward the end of the nineteenth century, many scientists were conducting research on the radiative and absorptive properties of gases in the atmosphere. Following in this trend, Svante Arrhenius presented a paper to the Royal Swedish Academy of Sciences in 1895 that showed a doubling of CO_2 would lead to a rise in global temperature of about 6.0°C (10.8°F); the paper was later published by the *Philosophical Magazine* (Arrhenius, 1896). By the 1930s, G. S. Callendar, of London's Imperial College of Science, had calculated the amount of CO_2 humans had emitted into the atmosphere. Callendar (1938) concluded that the observed rate of CO_2 increase could lead to a 1.1°C (2.0°F) warming per century. Several decades later, Johns Hopkins University scientist Gilbert Plass (1956) determined that a doubling of CO_2 would force the planetary temperature to rise by 3.6°C (6.5°F). By the early 1960s, Möller (1963) was developing very simplistic models of the atmosphere; his research led to an estimate of a 1.5°C (2.7°F) rise in temperature for a 300 ppm rise in atmospheric CO_2. The calculations were becoming increasingly intricate and complex by the end of the 1960s; the modern numerical models of global climate were becoming an important part of the CO_2 research. Nonetheless, the early "pioneering" work by Arrhenius, Callendar, Plass, and Möller led them to conclusions that are remarkably consistent with the predictions of some of the most complex climate models.

WHAT ARE NUMERICAL CLIMATE MODELS?

Models are idealized representations of reality, and in the context of numerical global climate models, they are mathematical, theoretical, deductive, and deterministic representations of the climate. The goal of the numerical modelers is to generate models that are based on the physics governing the mass, momentum, and energy flows and exchanges in the atmospheric system. For example, in Chapter 1 an equation was given for calculating the effective temperature of the earth. In a very simplistic form, that equation could be considered a numerical model of climate. The equation was obviously mathematical, it can be derived from relatively simple theory, and its derivation was deductive. We started with the theory, then built the equation, as opposed to measuring the solar constants and various effective temperatures of the planets and moons and then finding an equation that matched our observations. The effective temperature equation is also deterministic—the most basic forcing functions of global temperature are explicitly represented in the model. If we were to calculate the effective temperature for the various planets and moons of the solar system, and then compare the estimated effective temperature with the actual mean temperatures of these bodies, we would be close in many cases and quite in error in other cases. Nonetheless, we would still have a numerical model capable of simulating, with some limited degree of accuracy, the mean global temperature of bodies in the solar system.

The effective temperature "model," which is clearly at the lower end of the spectrum of model complexity, is an example of a zero-dimensional model (Schneider and Dickinson, 1974; Fraedrick, 1978).

It is called zero-dimensional because it does not resolve any of the latitudinal, longitudinal, or vertical patterns in the climate system. Given the extreme limitations in using such a zero-dimensional model, one may conclude that the only useful climate models must be three-dimensional. However, a number of one-dimensional models have proved useful in CO_2–climate research.

Imagine a vertical line running from the surface of the earth straight out to the very top of the atmosphere. At many points along the line, we could specify various physically based equations that could simulate the transfer of solar energy, the transfer of infrared energy from the earth and atmosphere, the vertical movement of air via convective processes, and even some basic cloud physics (e.g., Manabe and Wetherald, 1967; Manabe, 1983). The influence of various gases could be carefully specified in such a radiative-convective model, and as the concentrations of these gases are altered, the effects on energy transfers, temperatures, convection, and clouds could be determined. Such a one-dimensional model is surprisingly well suited to the greenhouse problem, and although its one-dimensional character would seem very limiting, these models have been used successfully in greenhouse research (e.g., Manabe and Wetherald, 1967; Schneider, 1975; Watts, 1980).

Another type of popular one-dimensional model resolves latitudinal differences in climate as opposed to the vertical structure of the atmosphere. Sellers (1969) and Budyko (1969) independently developed two of the most widely used one-dimensional energy balance models that have been applied to the greenhouse question. However, these models are largely used in classroom exercises, and have not continued to be utilized in many recent greenhouse experiments. Two-dimensional models (e.g., Sellers, 1973) combining a vertical coordinate with latitude or including only longitude and latitude are uncommon, and have not played a significant role in the greenhouse research.

Within the hierarchy of models (Schneider and Dickinson, 1974; Gal-Chen and Schneider, 1976), the three-dimensional models are clearly at the top, and these three-dimensional models are central to the greenhouse debate. These models attempt to resolve the latitudinal, longitudinal, and vertical components of the earth-atmosphere system. To visualize how many of these models operate, think about a grid of points over the earth's surface. Although the models vary in terms of spatial resolution, a grid of approximately 500 km by 500 km

(300 by 300 miles) is common. Even at the rather coarse spatial resolution of the 500-km squares, one should realize that several thousand of these squares are needed to cover the globe. Because the three-dimensional models contain a vertical component, these several thousand squares defined at the surface have layers of boxes above. Most of the modern three-dimensional models have approximately ten vertical layers, and therefore, the earth-atmosphere system is represented by over 20,000 boxes.

An intricate and complex set of equations is solved for each grid point or box to determine time and space changes in mass, energy, and momentum. The equations are written to carefully simulate changes in atmospheric pressure, fluxes of incoming solar energy, outgoing infrared radiant energy, thermal patterns, wind vectors, moisture levels, precipitation, clouds, ice and snow, and on and on. If there is not enough complexity already, the models should simulate oceanic circulations and allow a coupling between the oceans and the atmosphere. Because many of the three-dimensional models are based fundamentally upon the equations governing the wind patterns of the planet, these three-dimensional models are often referred to as general circulation models or GCMs.

All modelers are confronted with finding a balance between the physical representation of the climate elements and speed of computation (in fact, to maximize computational efficiency, some "spectral" models do not have grids and boxes, but rather produce all calculations for a series of harmonic waves). Ideally, modelers seek to represent all of the processes with theoretically based equations generated from the underlying physics. However, this goal is compromised at times to allow the computer program making up the model to run more quickly.

Many processes operating within the earth-atmosphere system can be represented with more simplified equations that are based on observed statistical relations. These simplified equations may have great accuracy in representing some process in the atmosphere, but they are not equations that reflect the physics of the process. These "fast physics" relations are referred to as parameterizations. They keep the computation time down, but the parameterizations reduce the scientific purity of the model. Many parameterizations used in the earlier models are fortunately being replaced by more explicit and physically based equations in the latest generation of climate models. Convective processes, heat flow in the soil, sea ice processes, and the structure of the cloud deck are examples of recent improvements in the models. However, sub-grid-scale phenomena, such as thunderstorms operating at a scale less than the 500 km grid spacing, continue to be parameterized in the models.

Imagine that the computer program is written and ready for a climate simulation. The surface conditions, including basic geography and topography, are specified along with starting conditions in the atmosphere; obviously, detailed information about the sun and the orbit of the earth can be specified in the model. The equations that make up the model are written in a form that allows the change in surface and atmospheric conditions to be calculated for a given change in time (time steps near 30 minutes are common). The models are started or initialized with the surface and atmospheric conditions, and all equations are solved for the change in the atmospheric and surface components over one time-step interval. This produces a new set of conditions, and the model equations are once again solved for another time step. After several years of simulated time in the model, the calculations stabilize, and outputs can be generated for a large number of simulated surface and atmospheric conditions (Meehl, 1984).

These models represent enormously complex computer programs that are tremendous achievements in computing, applied mathematics, and atmospheric physics. Many of the best minds in climatology have been used to construct these models, which require the power of the world's biggest and fastest computers. In the late 1960s and early 1970s, the climate models were typically constructed by just one or two investigators (e.g., Sellers, 1969; Budyko, 1969; Manabe and Wetherald, 1975). By the 1980s and 1990s, the models tended to be constructed by large groups of scientists.

The models most widely used in greenhouse research are often referred to by the organization, as opposed to the scientists, that built the model. Each of these organizations has produced a numerical model of the global climate; each model is different in resolution, time step, and process representation, and each model is likely to produce a different global climate response to a doubling of CO_2. Leading modeling groups include the Geophysical Fluid Dynamics Laboratory (GFDL) in New Jersey, the National Center for Atmospheric Research (NCAR) in Colorado, Oregon State University (OSU), the Goddard Institute for Space Studies (GISS) in New York, the Meteorological Office, United Kingdom (UKMO), the Canadian Climate Centre

(CCC) in Ontario, the Main Geophysical Observatory (MGO) in St. Petersburg, the Meteorological Research Institute (MRI) in Japan, the Commonwealth Scientific and Industrial Research Organization (CSIRO) in Australia, and the Max Planck Institut für Meteorologie (MPI) in Germany (Cubasch and Cess, 1990).

VERIFICATION OF THE MODELS

Several interrelated steps are required for a model to simulate climate conditions for a world with a doubled concentration of atmospheric CO_2 (or equivalent CO_2). The obvious first step is to run the model for present conditions, and verify that the model can fairly accurately simulate the observed climate. Often, these models are "tuned" to provide a more accurate simulation of the present condition. Tuning involves adjusting various terms in the models upward and downward until the model achieves some level of accuracy. Once the model has been tuned and is able to simulate present conditions fairly accurately, the model is then used in an experiment with increased CO_2. Virtually all CO_2 experiments start off with these modern-day simulations, and although the errors vary greatly among the many models, a few generalizations can be made (Gates et al., 1990; Kalkstein, 1991).

Recalling that the three-dimensional numerical models of climate are often called general circulation models due to the emphasis on simulating the earth's wind systems, we might expect the models to successfully replicate the observed circulation patterns. Generally, all three-dimensional models are capable of simulating the low-pressure trough extending around the planet near the equator, the high-pressure cells of the subtropical desert areas, and the general storm tracks of the mid-latitudes. They also generally capture the circulation changes associated with the shifting seasons. While these global-scale features are reasonably well represented, the regional-scale features such as the monsoon circulation of Southeast Asia are rather crudely simulated in the models.

As with the circulation features, the simulated temperatures of the planet are reasonably well represented at the global scale, but substantial errors occur at more local scales. The equator is warm in the models and the poles are cold, and the temperatures rise and fall with the seasons. However, at the scale of continents or subcontinents, the errors in the simulated temperatures are significant. For example, Table 1

Table 1. Errors in Eight Selected Model Simulations for Central North American Temperature and Precipitation.

Model	Winter Temp. (°C)	Winter Precip. (%)	Summer Temp. (°C)	Summer Precip. (%)
Canadian Climate Center	−1.4	+27	−0.2	+58
National Center for Atmospheric Research	+3.5	+45	+8.5	−58
GFDL (low resolution)	+1.3	+73	+4.5	+37
GFDL (high resolution)	−0.3	+18	+2.3	−12
Goddard Institute for Space Studies	+5.8	+82	−1.9	+29
Oregon State University	+2.2	+27	−1.0	−29
UKMO (low resolution)	+5.3	+9	−1.9	+67
UKMO (high resolution)	−4.4	−9	−1.2	+12

Note: GFDL is the Geophysical Fluid Dynamics Laboratory model and UKMO is the United Kingdom Meteorological Office model. All data are from Gates et al. (1990).

presents the mean errors in seasonal surface air temperatures for an area in central North America bounded by 35° to 50°N and 85° to 105°W (central New Mexico to southern Saskatchewan to central Ontario to northern Georgia). The mean error for eight different models in the winter (December–February) is 3.0°C (5.4°F) while the mean error in summer (June–August) is 2.7°C (4.9°F). Given the complexities of the climate system, most climatologists regard the model accuracy at this scale as a significant scientific achievement; however, mean errors of this magnitude seen in central North America and throughout virtually all other land areas of the globe, cannot be neglected in interpreting various greenhouse simulations.

The ability of models to simulate precipitation patterns is directly related to the ability to accurately simulate temperature and circulation features. Errors found in the temperature and circulation components of the model are compounded to produce even larger errors in the precipitation simulations. Global-scale precipitation patterns are simulated with reasonable accuracy, although the models are often too wet in the subtropics and too dry during the monsoon season in Southeast Asia. Table 1 shows the mean error in winter precipitation in central North America to be 36 percent and the mean error in summer precipitation to be 38 percent; errors around the globe are typically between 20 percent and 50 percent. As with temperatures, the errors in the precipitation levels, including the sign of the errors, vary greatly from one model to the next.

Errors in circulation features, temperatures, and precipitation patterns are relatively easy to quantify given the available data bases for the earth's climate. However, errors in other variables such as soil moisture, snow cover, sea ice, and cloud patterns are more difficult to quantify given the general shortage of observational data.

SIMULATED RESPONSE TO A DOUBLING
OF EQUIVALENT CO_2

As we have seen, a number of sophisticated three-dimensional numerical climate models have been constructed based largely on the laws of physics that govern the earth-atmosphere system. And generally speaking, these models are capable of simulating the major features of the climate system. In order to conduct a doubled CO_2 experiment, scientists must alter the instructions to indicate to the computer program that the atmospheric concentration of CO_2 (or equivalent CO_2) has been doubled; typically, the value of CO_2 goes from 300 ppm to 600 ppm in the numerical experiment. The routines in the models that handle the radiative transfers recognize this alteration to the atmospheric chemistry, and they initiate a series of changes to the planetary climate system. In what are called equilibrium climate change experiments, the climate model is run for 300 ppm, then run for 600 ppm, and the differences in the simulated climate are determined. In more difficult, but more realistic time-dependent experiments, the model's equivalent CO_2 levels are gradually raised from 300 ppm to the 600 ppm level and the resultant climate is determined along the time continuum. Again, the climate calculated for the 300 ppm condition is subtracted from the climate simulated for the 600 ppm concentration and the differences are noted. These are somewhat different approaches to the problem (Schneider and Thompson, 1981), but they tend to produce similar results (Bryan et al., 1982; Spelman and Manabe, 1984; Meehl, 1984; Bretherton et al., 1990).

Virtually all climate models, from the zero-dimensional models to the most complex three-dimensional time-dependent models, predict an increase in global temperatures as a doubling of CO_2 occurs. Even in the last century, calculations were made predicting a rise in global temperature with a doubling of CO_2 (Arrhenius, 1896). And basically, climatologists have been arguing ever since about the magnitude, and not the possibility, of the resultant warming. As can be seen in Table 2,

Table 2.   Changes in Global Temperature (ΔT) and Precipitation (ΔP) in Eight Selected Models for a Doubling of CO_2.

Model	ΔT (°C)	ΔP (%)
Canadian Climate Center	3.5	4
National Center for Atmospheric Research	3.5–4.0	7–8
GFDL (low resolution)	2.0–4.0	3.5–9
GFDL (high resolution)	4.0	8
Goddard Institute for Space Studies	3.9–4.8	11–13
Oregon State University	2.8–4.4	8–11
UKMO (low resolution)	1.9–5.2	3–15
UKMO (high resolution)	3.5	9

Note: GFDL is the Geophysical Fluid Dynamics Laboratory model and UKMO is the United Kingdom Meteorological Office model. All data are from Cubasch and Cess (1990).

the range of projected global air temperature changes from selected recent three-dimensional climate models is from less than 2.0°C (3.6°F) to over 5.0°C (9.0°F). Recently, Schlesinger and Jiang (1991) concluded that 1.2°C (2.2°F) may be the expected rise in global temperature associated with a doubling of the greenhouse gases. Others (e.g., Lindzen, 1990) have suggested that improvements in the models will lead ultimately to a predicted rise in temperature of less than 1.0°C (1.8°F) as the concentration of greenhouse gases is doubled.

The geography and seasonality of the warming is fairly consistent from one model to the next. The tropics are predicted to warm the least while the high latitudes are predicted to warm the most; often the high latitudes are predicted to warm more than twice as much as the mean global warming. In addition, mid-latitude and high-latitude locations have far more warming in winter than in summer; the ratio of winter to summer warming is near 1 at about 30°N or S and over 5 in the highest latitudes. As an example, Figure 11 displays the projected latitudinal December–February warming for four different climate models. The plot clearly shows that the greatest amount of the warming occurs in the high latitudes during the winter season.

Just as every climate model predicts increasing temperature for a doubling of CO_2, every climate model also predicts a CO_2-induced increase in globally averaged precipitation. Figure 12 shows the positive, nearly linear relation between the predicted global warming and predicted increases in precipitation for 19 different model simulations. Global warming increases the global evaporation rates, the hydrological

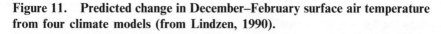

Figure 11.   Predicted change in December–February surface air temperature from four climate models (from Lindzen, 1990).

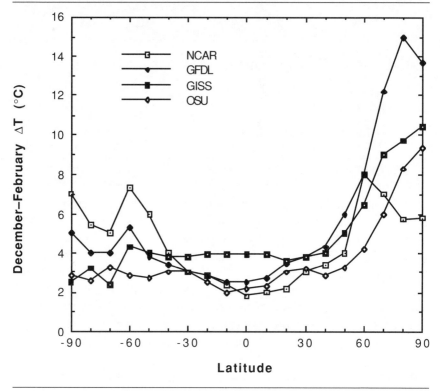

cycle is intensified, and more precipitation results. Although the results vary substantially from model to model, precipitation increases are predicted to be greatest in the mid to high latitudes, particularly in the stormy winter season. In many models, precipitation tends to decrease or remain unchanged during the summer season over the mid-latitude continents of the Northern Hemisphere. Although the "popular vision" suggests increased drought in many agricultural areas, the reality is that all climate models are predicting more annual precipitation to fall in almost all areas of the world.

The predicted increasing temperatures of the planet would certainly increase the rate of evaporation and transpiration (water entering the atmosphere through the leaves of plants). Obviously, the future evapo-transpiration rates will also depend on changes in clouds, incoming

Figure 12.   Predicted changes in global precipitation vs. changes in global temperature for a doubling of CO_2. Data for the 19 model runs are from Cubasch and Cess (1990).

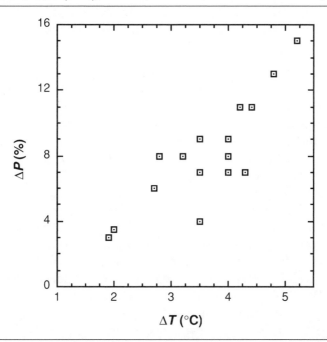

solar energy, humidity levels, and wind speeds, but any increase in temperature is very likely to create a rise in the evapotranspiration rates. The increase in evapotranspiration is predicted to overwhelm any increases in precipitation, and soil moisture levels are expected to decrease over many continental areas. The predictions of increased drought associated with a doubling of CO_2 stem largely from the temperature-driven increase in evapotranspiration, and not from any decline in precipitation. Rather clearly, any decline in mid-latitude summer precipitation over the continents could lead to devastating droughts if significant warming takes place (e.g., Manabe et al., 1981; Manabe and Wetherald, 1986, 1987; Gleick, 1987; Rind et al., 1990; McCabe et al., 1990; Mitchell et al., 1990). In addition, the climate models tend to predict that a doubling of CO_2 will reduce the diurnal temperature range (the difference between daily maximum and minimum temperatures) by a small amount, and decrease the area covered by snow and sea ice. According to the models, we are in for some significant changes in our climate, and these changes are going to occur very rapidly over the next half-century.

PROBLEMS AND LIMITATIONS OF THE MODELS

As we have seen, the many climate models of the world are suggesting substantial increases in annual temperature and precipitation as the concentration of CO_2 (or equivalent CO_2) moves from 300 ppm to 600 ppm. Without the predictions of the models, the greenhouse effect would never have become such a substantial scientific and popular issue. Many scientists, politicians, and decisionmakers have accepted these predictions as gospel—but too often, these predictions are presented without much discussion of the problems and limitations of the models.

Just as there is general agreement in the predicted rise in temperature and precipitation, there is also agreement that the models are far from perfect representations of reality. By definition, even the most sophisticated time-dependent three-dimensional general circulation models are highly simplified, extremely crude models of the climate system. While the numerical climate models will continue to be improved into the future, the existing simulations for a doubling of CO_2 are limited by the following shortcomings.

First, most climatologists agree that the key to understanding global climate change lies in the understanding of ocean-atmosphere interactions. The role of the ocean in absorbing CO_2, storing and transporting heat, and even releasing sea salts into the atmosphere is not well known and certainly not included adequately in the existing models. Much of the debate about the greenhouse effect centers on these ocean-related issues. Some scientists argue that due to the large thermal inertia of the oceans, any greenhouse warming will be delayed by 10 to 100 years (e.g., Ramanathan, 1981; Wigley and Schlesinger, 1985; Wilson and Mitchell, 1987; MacCracken, 1987; Schlesinger and Jiang, 1990). The climate consequence of our increase in the greenhouse gases may not be fully revealed for another century! However, others have argued that the thermal lag of the oceans is likely to be less than a decade, and possibly no more than a few years (Ellsaesser, 1990; Michaels, 1990). Until we have better knowledge of the coupling between the oceans and the atmosphere, the model predictions must be treated with considerable caution.

Second, it is widely recognized that the models do not adequately simulate the potential feedbacks associated with clouds. As the CO_2 doubles and the world presumably warms, changes are likely to occur in cloud types, areal coverage of clouds, droplet size, liquid water content, and cloud albedo. In some cases, it is not clear whether given changes in cloud properties will act to warm or cool the earth. If the greenhouse world produces more low, thick, stratus clouds, the clouds should act to cool the earth; however, an increase in the high, thin cirrus clouds may act to warm the earth. Recent evidence shows that the net effect of increasing cloud cover would be a net cooling of the planet (Hartmann and Doelling, 1991); however, changes in cloud types could complicate the response of the global temperature. The numerical climate models vary widely in their cloud-climate responses, and until significant improvements are made, many important questions concerning the overall climate response to doubling CO_2 will remain unanswered (e.g., Paltridge, 1980; Wetherald and Manabe, 1986; Cess et al., 1989; Mitchell et al., 1989; Ramanathan et al., 1989b; Lindzen, 1990; Del Genio et al., 1991; Ramanathan and Collins, 1991; Schmitt and Randall, 1991).

Arguably, ocean-related and cloud-related problems represent the most significant shortcomings in the models. Other widely acknowledged problems in the models are related to sea ice and snow cover, representation of biological systems, and the role of other pollutants that may act to cool, rather than warm, the earth (see Dickinson, 1989 for an excellent overview). Sub-grid-scale phenomenon such as convective storms are still not well represented in even the best numerical climate models (Stone and Risbey, 1990). In some cases the modelers are limited by computing power, and in other cases the modelers are limited by our poor understanding of the physics of the atmosphere.

What we are left with are sophisticated models that have been constructed to conduct a variety of numerical climate experiments. It is important to realize that the models were not built just for conducting the now-important $2\times CO_2$ simulations. The models are able to simulate the gross features of the climate system, but they are not capable of refining the details of the regional variations in climate. On one hand, they are extremely complex and notable scientific achievements, but on the other hand, they are still extremely crude representations of reality. The models will improve in many aspects in the near future, and the predicted climate response to a doubling of CO_2 is likely to change as well.

Decisionmakers must be aware of the severe limitations of the models. While these models may provide some useful information on global-scale climate changes, both the American Meteorological Society and the Royal Meteorological Society have recently concluded in their respective policy statements that the models are incapable of providing detailed predictions at highly localized geographic scales (AMS, 1991; RMS, 1991). Obviously, if policies regarding global change are going to be based on the predictions of the models, it is absolutely imperative that the policymakers appreciate the strengths, weaknesses, and limitations of the existing general circulation models.

4

OBSERVED GLOBAL-SCALE
CLIMATE CHANGE

Virtually every greenhouse scientist would agree on two points: the models are generally predicting a rise in temperature of between 1.0°C (1.8°F) and 5.0°C (9.0°F) for a doubling of equivalent CO_2, and equivalent CO_2 levels have risen at an exponential rate over the past century. Many are sufficiently convinced by these two facts to embrace global warming as a very real potential threat to the planet that must be dealt with immediately. Other scientists are concerned about the present limitations of the models, and they have placed their energies in constructing the next generation of improved numerical climate models. However, another group of climatologists has approached the problem somewhat differently. These are the empiricists; these scientists are seeking to identify a linkage between observed changes in climate and observed changes in atmospheric chemistry. Much of the research done by the empiricists leads one to conclude that the greenhouse effect may not produce the catastrophic changes predicted by the numerical climate models. These models may be off by a factor of two, if not more!

In almost every case, the greenhouse argument is hinged on the calculation of what happens when the equivalent CO_2 goes from 300 to 600 ppm. The simulated climates for the two conditions are subtracted from one another, and the difference is touted as the expected climate response to a doubling of CO_2. Looking back on Figure 10,

47

we find that the equivalent CO_2 rose from 310 ppm near the turn of the century to over 430 ppm by 1990. Over the past 100 years, equivalent CO_2 levels have already increased by 40 percent! Empiricists argue that we should use the climate models to simulate the changes in CO_2 over the past 100 years and generate an expectation of global warming for that period. Given some estimate of global temperature over that same 100-year period, we could do an experiment to check the accuracy of the numerical climate models in predicting climate response to increasing CO_2.

A number of climatologists have discussed extensively the issue of predicted climate response for the changes in equivalent CO_2 observed over the past 100 years. Generally speaking, the models predict a rise in global temperature between 0.5°C (0.9°F) and 2.0°C (3.6°F) for the change in greenhouse gases of the past century (Lindzen, 1990). More specifically, MacCracken (1987) estimated that between 1.1°C (2.0°F) and 1.3°C (2.3°F) of greenhouse warming should have taken place since the 1850s, while Schneider (1989a) listed 1.0°C (1.8°F) of expected warming over the past 100 years. Michaels (1990) argued that the existing models imply a greenhouse warming for the last 100 years to be near 1.7°C (3.1°F); he suggests that even the most liberal estimates of the ocean thermal lag (which delays global warming) still leaves the expected warming for the last century to be between 1.0°C (1.8°F) and 1.2°C (2.2°F). Wigley and Barnett (1990) also concluded that the expected greenhouse signal of the past 100 years should be between 1.0°C (1.8°F) and 2.0°C (3.6°F). Jones and Wigley (1990a) suggested that the buildup of greenhouse gases over the last century will eventually cause a 0.8°C (1.4°F) to 2.6°C (4.7°F) increase in global temperature. However, due to the ocean thermal lag effect, they argued that we should have witnessed only 0.5°C (0.9°F) to 1.3°C (2.3°F) of global warming. In this chapter, we will explore whether or not the planetary climate system of the past 100 years has produced any signals that are consistent with the expectation of the models.

LONG-TERM VARIATIONS IN GLOBAL CLIMATE

Before we explore the nature of climate variations over the past century, it is critical to note a few features about longer term climate changes. Figure 1 shows that the planetary temperature has always been changing; in fact, climate change is clearly the rule and not the exception. This figure shows that over the past 850,000 years (and recall that the earth is approximately 5 billion years old), the planetary temperature has shown tremendous variability with a range of more than 6.5°C (11.7°F). The pattern is one of rather abrupt jumps in temperature—there is little evidence of climate stability in the past, and it is not logical to expect the modern-day climate to remain stagnant, with or without the addition of the greenhouse gases. The problem here is rather simple: climatologists do not fully understand the reasons for the variations of the past, and the scientists are not sure where the climate of the twentieth century was headed in the absence of the buildup of the greenhouse gases. Given these unknowns, determination of a pure greenhouse signal in the global climatic data is made very difficult, if not impossible.

Climate variations at the time scale of the last 1 million years are probably linked closely to changes in the earth's orbit around the sun, change in the sun's output, random fluctuations in the climate system, volcanic activity, and preindustrial variations in carbon dioxide. Indeed, many investigators have noted a rather close and direct relationship between the CO_2 levels measured in the Vostok ice core (Figure 2) and the planetary temperature (Figure 1). Some scientists (e.g., Genthon et al., 1987; Tangley, 1988) have suggested that the Vostok ice record confirms the importance of CO_2 in determining a significant portion of the variation in the planetary temperature. Many other scientists (see Idso, 1989, for a review) have argued that at this long time scale, the changes in climate may actually be the cause of the observed changes in CO_2. The earth's orbit around the sun is not constant, and variations in the orbit explain some of the temperature patterns of the past million years (see Sellers, 1965, for a review). The eruption of large volcanoes can inject dust into the stratosphere and ultimately act to cool the planet. Some scientists (e.g., Bryson, 1989) have been able to link periods of volcanism to long-term climate change, but others (see Sellers, 1965) have found problems with the explanation. Still other scientists (Lorenz, 1970; Houghton et al., 1991; Gordon, 1991) have argued that even in the absence of some external influence (e.g., volcanism, variations in earth-sun geometry), the climate system would show random fluctuations that could lead to many of the trends and variations found in the long-term temperature record. This entire book could be devoted to the causes of long-term climate changes, but the bottom line is that scientists are not sure where the natural drift of climate was headed in the twentieth century.

Approximately 500 years ago, much of the Northern Hemisphere plunged into a particularly cool period known as the Little Ice Age (Grove, 1988). This period produced mean global temperatures that were nearly 1.0°C (1.8°F) below the modern-day normals, and the impacts of the cooling on the social and economic systems of that time in human history have been well documented (see Lamb, 1982). No significant drop in atmospheric CO_2 appeared at the beginning of the Little Ice Age, so CO_2 was not the culprit. Instead, increases in volcanic activity, a reduction in solar output, and/or changes in ocean currents most likely sent the planetary temperatures falling (Bryson, 1989; Idso, 1989). Substantial warming about 100 years ago ended the Little Ice Age, and this warming coincides nicely with the buildup of anthropogenic CO_2. However, it remains largely unknown whether any warming this century is predominantly from emissions of greenhouse gases or from some processes that brought us rather naturally out of the Little Ice Age.

As we explore the changes to the earth's temperature over the past century, we must remember that any observed warming (or cooling) is not necessarily caused by the increasing concentration of CO_2. A multitude of processes are at work on the climate system, and we know that these processes have warmed and cooled the earth repeatedly in the past. There is absolutely no guarantee that any CO_2–climate signal over the past century can be identified in a system that has such a high level of natural variability.

HISTORICAL PATTERNS IN GLOBAL TEMPERATURES

Currently, the most widely used global surface air temperature data base is a time series developed by the Climatic Research Unit at the University of East Anglia in Norwich, England. Philip Jones and his colleagues carefully selected meteorological stations and fixed-position weather ship records from throughout the globe (Jones et al., 1986a, 1986b, 1986c). His group screened the temperature data to eliminate as many nonclimatic errors as possible (e.g., station moves, instrument changes). Efforts were made to avoid major urban centers where urban-induced climate changes could contaminate the record. Jones began with 2,666 stations in the Northern Hemisphere and 610 stations in the Southern Hemisphere. After carefully screening the data and adjusting some of the observations, Jones and his co-workers ended with 1,584 and 293 stations in the two hemispheres, respectively. The oceanic temperatures were extracted from literally millions of ship observations of sea-surface and air temperatures. As with the land-based records, extensive quality control analyses were performed on the ocean records.

A 5° latitude by 10° longitude grid was established, the station and ocean records were converted to monthly temperature anomalies (actual monthly temperature minus the normal monthly temperature), and the resulting temperature anomalies were then interpolated to the grid points. This produced a monthly temperature value for each grid point; the period of record for many points extends from 1861 to the present.

Obviously, the number of available station and ship reports increases over the period 1861 to present. The number of reporting locations before the 1880s was relatively low, and the reliability of the records is in question. Schönwiese (1987), among others, cautioned against the use of the earliest of these records because of the lower number of stations and the uncertainties in the collection program. Jones and Wigley (1990a) concluded that the temperatures before 1880 are only half as accurate as the temperature records after 1920.

Figure 13 shows the global temperature plot over the period 1891 to 1990 along with a least-squares line showing the linear trend over the 100 years of record. The temperatures for the planet appear to increase from 1891 to about 1940, decrease slightly from 1940 to the late 1970s, then increase sharply in the 1980s. The slope of the line is 0.0045°C (0.0081°F) per year, representing a century-long linear increase in global temperature of 0.45°C (0.81°F). Given the scatter around the trend line, there exists a statistical uncertainty regarding the actual value of the trend over the past century. The underlying trend could be higher or lower than the computed value of 0.45°C (0.81°F); greater scatter around the trend line would lead to an increased uncertainty about the actual value of the underlying trend in the data. Fortunately, we can compute confidence intervals for the trend, and in the case of the 100-year linear trend, we can conclude with 95 percent statistical confidence that the linear trend is actually 0.0045°C (0.0081°F) per year plus or minus 0.0008°C (0.0014°F) per year. Therefore, we can say with 95 percent confidence that the 100-year trend *in this set of global temperature data* lies somewhere between 0.37°C (0.67°F) and 0.53°C (0.95°F).

Figure 13.   Global temperature anomalies (based on 1950–1979 normals) for the period 1891–1990. Data preparation is described by Jones et al. (1986c); actual data are available in Boden et al. (1990).

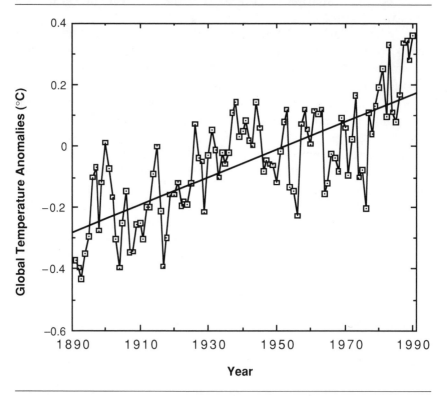

Several investigators (e.g., Karl et al., 1984; Folland et al., 1990) have shown that this rise in temperature has not been consistent through the entire 24-hour period. Evidence from a number of Northern Hemispheric stations suggests that nighttime temperatures are rising while daytime temperatures are falling or remaining unchanged. This reduction in the diurnal temperature range (the difference between maximum and minimum temperatures) is somewhat consistent with the predictions of some of the $2 \times CO_2$ model simulations (see Rind et al., 1989), although the observed decline in diurnal temperature range is substantially larger than what has been predicted by the models.

Herein lies a significant component in the greenhouse debate. As described earlier, the climate models suggest that we should have observed at least a 1.0°C (1.8°F) warming over the past century, even when the effects of oceanic thermal lags are considered. And yet, the best data set of its kind reveals a trend of only 0.45°C (0.81°F) over this time period. With only this level of analysis, one could conclude that the models must be off by at least a factor of two!

Recognizing the problem, noted modeler Stephen Schneider (1989a) of the National Center for Atmospheric Research (NCAR) offered the following possible explanations for the apparent discrepancy. First, the models have overestimated trace gas increases by about twice the actual amount. This explanation seems highly unlikely given our general knowledge of changes in atmospheric chemistry over the past century, and given the quality of scientists working in the modeling centers. Second, competitive external forcing (e.g., volcanic dust, solar output, effect of other pollutants) has not been properly accounted for by the models. This seems very possible given the fact that present model simulations of $2\times CO_2$ do not yet include volcanic dust, effects of other pollutants, or variations in solar output. Future simulations will undoubtedly include these effects, and very likely change the predictions for the climate of the next century. Third, the large heat capacity of the ocean has not been properly accounted for by the existing models. But as we saw earlier, many scientists have concluded that even with the ocean thermal lags accounted for, the amount of warming over the past 100 years should still be near 1.0°C (1.8°F). Fourth, Schneider argued that models have been run for a doubling of equivalent CO_2, and not for the 25 percent increase actually experienced. While it is true that the models have been run almost exclusively for a doubling of CO_2, most estimates of the rise in equivalent CO_2 over the last century are closer to 40 percent than to 25 percent (see Figure 10). Finally, Schneider suggested that the incomplete and nonuniform network of thermometers have underestimated actual global warming in this century. As we shall see, the station network probably has overestimated and not underestimated global warming over the past century. Schneider (1989a, p. 118) concluded, "Despite the litany of potential excuses, the twofold discrepancy between GCM predictions and observed global temperature trends is still fairly small."

The problem does not go away so easily. Suggesting that a twofold discrepancy is fairly small obviously is a matter of one's perspective on the greenhouse issue. But to conclude that the discrepancy is only twofold assumes that (1) the amount of global warming is at least

0.5°C (0.9°F), (2) all of the observed warming can be attributed to the buildup of the greenhouse gases, and (3) most of the warming must occur after, not before, the large increase in atmospheric concentration of the trace gases. Three highly significant facts must be considered when examining these assumptions and the mismatch between the observed change in global temperatures and the expected change predicted by the models. First, the global temperature record contains a number of biases (e.g., urban heat islands, measurement problems) that tend to produce warming signals not related to the buildup of greenhouse gases. Second, as suggested by Schneider, other external forcings (e.g., volcanic eruptions) absolutely account for some of the trend in global temperatures, leaving even *less* global warming unexplained. And third, much of the warming of the past century took place before the end of World War II—the timing of the warming does not relate well to the increase in equivalent CO_2.

CONTAMINANTS TO THE TEMPERATURE RECORD

At first glance the calculation of the trend in planetary temperature would seem rather easy. The Jones data base has been established for thousands of land-based stations and millions of ship observations, the data have been carefully screened for reliability, and these data have been gridded and averaged for the globe. Rather simple statistical analyses, as well as very sophisticated ones, produce a trend of approximately 0.45°C (0.81°F). However, the problem of determining a trend in the planet's temperature is not so simple. The temperature records of the globe contain any number of non-greenhouse-related signals that jeopardize the reliability of the long-term trend in temperature.

Reliability of the Temperature Record

A variety of problems stem from the actual observation of the atmospheric temperature. Obviously, no one observer can take the readings over the course of an entire century. Therefore, as the chore of taking the daily observations passes from one observer to the next, the instruments are likely to be moved to a new location, probably not far from the previous observation point. While even the shortest of moves may seem trivial, changes in exposure, elevation, and topography can change the recorded temperature and create a discontinuity in the record (see Mitchell, 1953; Karl and Williams, 1987; Karl et al., 1989). If the station move is well documented, some of the effects of the relocation can be statistically removed from the record.

The instruments themselves, along with the recommended exposure to the sun, have changed through time. In many parts of the world, observers are being replaced with computerized electronic sensors. These new sensors may be more accurate in measuring air temperatures at short time intervals, but how the readings from the electronic sensors compare to the readings from the traditional manually read maximum and minimum thermometers is still very much in doubt (Karl and Quayle, 1988).

In addition to potential shifts in the station location and changes in the instruments, the time of observation may change from one observer to the next. To understand this problem, imagine an observer taking readings on a bitterly cold winter morning. The observer records the maximum and minimum temperature over the past 24 hours, and then resets the equipment. Immediately, the bitterly cold temperature of that morning becomes the low temperature for the next 24-hour period. Even if a substantial warming occurs over the next 24 hours, the low temperature for that bitterly cold morning will be the established low. In essence, morning observers have every opportunity to double-count low temperatures. Conversely, afternoon observers would tend to double-count extreme high temperatures; an afternoon observer records the extreme maximum temperature, resets the equipment, and immediately has double-counted the high value. As the once-a-day, manually read systems are replaced by computerized equipment capable of 24 observations per day, the long-term record is impacted by a time-of-observation bias. Fortunately, if these details are well documented, an adjustment can be made to the temperature record and some of the time-of-observation biases can be removed (see Karl et al., 1986b).

In some areas of the globe, the mean daily temperature is calculated as the average of the daily maximum and minimum temperature. Other places use the average of the 24 hourly observations, while other places use the average of measurements made at 8-, 6-, or 4-hour intervals. As with the time-of-observation bias, the method used to calculate the daily and monthly mean temperatures can significantly affect the resultant value. Obviously, any changes in the averaging scheme over time can produce another bias in the long-term record.

The climate records from ships are biased by any number of reliability problems. Before 1940, sea-surface temperatures were taken by sailors pulling up seawater in buckets, then taking its temperature within the buckets on the deck of the ship. Some ships used wooden buckets while others used canvas buckets; after World War II, some ships shifted to plastic buckets. A number of experiments have shown that the type of bucket used can significantly affect the resultant sea-surface temperature. For example, from 1900 to about 1930, canvas buckets were not insulated, and evaporative cooling may have provided a cool bias. After 1940, most sea-surface temperatures were taken within the intake pipes of the ship's cooling system; unfortunately, these intake measurements are generally about 0.4°C (0.7°F) warmer than the canvas bucket readings. Obviously, every effort must be made to correct the ocean temperature records for the method used to measure the temperature of the seawater (see Jones and Wigley, 1990a).

The marine air temperature measurements are also prone to discontinuities through time. Possibly the greatest problem is that the ships of the world are getting larger, and the thermometers used to measure the air temperature are, on average, getting higher above the ocean surface. This change in height, along with many changes in observation time, exposure to the sun, and proximity to potentially warm features on the ship make the marine air temperature records particularly difficult to adjust to some baseline level.

All of these problems, along with many others, influence the reliability of the long-term climate record to some degree, and despite every effort to remove or minimize their effect, the record remains contaminated with these uncertainties. In addition to these measurement problems, the geographic distribution of the stations and ship records creates yet another difficulty for the "global" temperature record. Some areas of the world are well sampled with the existing network, while other areas are virtually unmeasured. The use of the grid by Jones et al. (1986c) helps to minimize the effects of uneven sampling over the globe, but nonetheless, the sparsity of records from major portions of the planet detracts from the calculation of a true global temperature. While some of these problems in the record will tend to cancel out, they absolutely increase the uncertainty in the 0.45°C (0.81°F) trend of the past 100 years.

The Urban Heat Island

The search for the global warming signal is complicated by the contaminants to the historical temperature records described above. However, while each of the problems described above cannot be overlooked in the search for the greenhouse signal, the potential impact on the temperature record caused by the urban heat island effect may dwarf them all. Recognizing the potential magnitude of this problem, a number of scientists have attempted to explicitly quantify the urban heat island effect in the historical land-based temperature record.

The physical causes for localized warming in urban areas have been studied vigorously for more than a century (e.g., Howard, 1833), and these causes are reasonably well understood (excellent review articles include Oke, 1979; Landsberg, 1981; Lee, 1984; Kukla et al., 1986; Brazel, 1987; Karl et al., 1988). Generally, the principal reason for the development of the heat island is the waterproofing of the urban surface. The natural vegetation is largely removed within cities, and the resultant surface is covered by hard, somewhat impervious materials. Rainfall quickly flows off the urban surface into underground storm sewers, and the resulting surface and near-surface moisture is minimized. The sunlight finds little water to evaporate, and therefore its energy goes to heating the ground and air as opposed to evaporating water; the result is a substantial increase in local temperatures.

A second major reason for the development of the heat islands is the lower reflectivity of the urban landscape. The lower reflectivity causes more of the sun's energy to be absorbed at the surface, and the surface warms more quickly. The complex geometry of the buildings within the city can create a greater absorption of incoming shortwave radiation due to the variety of exposures relative to the rays of the sun. The actual thermal properties of building materials may also contribute to the increase in local temperatures. For example, some building materials store more heat during the daytime than would be stored by the naturally occurring dry soils. Once this energy is absorbed, the urban canyons reduce the loss of long-wave infrared radiation by narrowing the view of the cold sky. Waste heat energy escaping into the free atmosphere also contributes to warming, particularly in the winter season in mid and high latitude locations. These processes, along with many others, are largely responsible for the development of the urban heat island effect.

A good example of the heat island effect on long-term temperature records is Phoenix, Arizona. The Phoenix metropolitan area has grown from just over 100,000 residents in the early 1930s to over 2 million in 1990. The expansion of the city into once irrigated areas, the large amount of sunlight of the desert setting, and the light winds of the area all produce a classic heat island (see Balling and Brazel, 1986a).

A plot of Phoenix's mean annual temperatures (Figure 14) shows a linear rise of 2.5°C (4.5°F) over the entire 1931–1990 time period. However, the linear rise in temperature from 1960–1990 is an incredible 3.8°C (6.8°F). Temperatures from surrounding rural stations in Arizona show a warming of less than 0.5°C (0.9°F) over the recent period

Figure 14.   Mean annual temperature in Phoenix, Arizona, from 1931 to 1990 along with population estimates for the Phoenix metropolitan area.

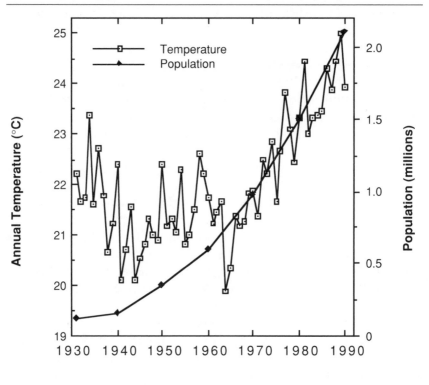

(Balling and Idso, 1989). Maps of the Phoenix area produced with temperature records from around the city and from satellite-based sensor systems (see Balling and Brazel, 1987, 1988, 1989) show the heat island effects to be confined to the metropolitan area. The plot of population levels in Figure 14 strongly suggests that the changes in Phoenix are far more related to the population growth than to some greenhouse effect.

The example of Phoenix vividly illustrates how growth in a city can create a substantial rise in local temperature. Recognizing the urban growth in many parts of the world, some climatologists have attempted to extract the global heat island signal from any warming that may be resulting from the buildup of greenhouse gases. Fortunately, researchers are approaching an agreement on the question.

The widely used Jones et al. (1986c) gridded temperature data base has been analyzed carefully to identify the heat-island signal in the global temperature record. One such analysis has been conducted by Thomas Karl and his associates at the National Climatic Data Center in Asheville, North Carolina. They have been examining long-term records from throughout the conterminous United States. They have selected mostly rural stations, carefully studied the many station histories, and statistically removed the effects of time-of-observation biases, station and instrument changes, and any lingering urban heat island effects. The result of their enormous effort is the United States Historical Climatology Network (HCN) data base (Quinlan et al., 1987), considered by many climatologists to be the best climatic data set of its type.

By comparing the gridded data for the United States from the Jones et al. temperature record to the temperatures from the HCN, Jones et al. (1989) computed an urban bias in their temperature data of 0.08°C (0.14°F) over the period 1901–1984. Arguing that no more than 40 percent of the Northern Hemisphere landmass is likely to have a similar bias, they concluded that the level of heat island bias in the Jones et al. (1986a) temperatures for the Northern Hemisphere is probably between 0.01°C (0.02°F) and 0.10°C (0.18°F); a value near 0.05°C (0.09°F) was considered most reasonable. Karl et al. (1988) repeated the analysis, using a different method, and found a 0.11°C (0.20°F) urban effect in the same data. Jones et al. (1990) continued with these kinds of analyses and found a heat island bias of 0.15°C (0.27°F) for the United States over the period 1901–1984. Jones et al. (1990)

further studied western U.S.S.R., eastern Australia, and eastern China and ultimately concluded that the global-scale urban bias is probably between 0.01°C (0.02°F) and 0.10°C (0.18°F), with the most likely estimate near 0.05°C (0.09°F) for the entire planet for this century. They noted that this value is fully an order of magnitude less than the observed temperature trend over the same time period.

When compared to an observed rise in temperature of 0.45°C (0.81°F), the 0.05°C (0.09°F) urban heat island signal may at first glance seem almost trivial. Nonetheless, this bias represents more than 10 percent of the total observed trend, and if Jones et al.'s (1990) high-end estimate of 0.10°C (0.18°F) is correct, the urban heat island bias represents nearly 25 percent of the global warming of the past century.

The debate regarding urban heat islands is likely to continue into the immediate future. However, at this time one could conclude that the Jones et al. (1986c) data set has a global urban warming bias somewhere between 0.01°C (0.02°F) and 0.10°C (0.18°F), with the most likely value near 0.05°C (0.09°F). The recent data set introduced by Vinnikov et al. (1990) appears to be similar to the Jones et al. (1986c) data, but the global data set developed and used by Hansen and Lebedeff (1987) appears to be much more seriously affected by heat island contamination.

Desertification and Global Warming

The section on urban heat islands showed that some of the global temperature rise of the past 100 years can be ascribed to the warming associated with the buildup of cities. The urban effect creates a localized warming signal that is not representative of the surrounding area. Recently, it has been discovered that overgrazing and desertification may be producing a large-scale warming signal that also is clearly not related to the greenhouse gases.

The role of desertification in changing the regional temperature was strongly debated following a landmark article by Jule Charney (1975) of the Massachusetts Institute of Technology. Charney suggested that overgrazing in arid and semiarid lands would increase the albedo by removing the dark vegetation, the increased albedo would reflect more of the sun's energy, less solar energy would be absorbed by the surface, and surface and air temperatures would drop. The cooler temperatures might reduce the potential for summertime convective storms, and the local rainfall would be diminished. The lack of rainfall would further stress remaining plants, and the resulting positive feedbacks would act to accelerate the desertification process. Soon after the introduction of the Charney hypothesis, Jackson and Idso (1975) and others argued that removal of vegetation would reduce evapotranspiration rates, less solar energy would be consumed in evaporating and transpiring water, leaving more solar energy to warm the surface and the air. Most empirical data appeared to support the notion that overgrazing and desertification would act to warm, not cool, the surface and air temperatures.

A few locations can serve to test whether overgrazing and desertification create local warming or cooling. One such location is along the United States–Mexico border in southern Arizona and northern Sonora. Following the passage of the Taylor Grazing Act in 1934, the rangelands on the American side of the border were largely protected from severe overgrazing. However, for any number of social, economic, and political reasons, the Mexican landscape has been severely overgrazed. Today, the fence marking the international border also defines a sharp discontinuity in surface conditions (see Figure 15). When compared to the adjacent landscape in Mexico, the lands in the United States have 28 percent more grass cover, 29 percent less bare soil, grasses that are 66 percent taller, and a 5 percent lower albedo (see Balling, 1988).

Recognizing the opportunity to examine the effects of this discontinuity on local climate conditions, researchers at the Jet Propulsion Laboratory in Pasadena, California, teamed with scientists at the University of Arizona and Arizona State University; I was a team member. We initiated a multiyear study to determine not only the magnitude of climate difference along the border, but also the physical causes of the differences. For the summer months, we found that the heavily overgrazed areas in northern Mexico had 2.5°C (4.5°F) to 4.0°C (7.2°F) higher air temperatures, up to 7.0°C (12.6°F) higher surface temperatures, and more than 15 percent higher potential evapotranspiration rates (Bryant et al., 1990). The overgrazing of northern Mexico has removed much of the plant cover, thereby accelerating the runoff of any local precipitation and leaving the soils relatively dry. The lack of soil moisture increases the surface temperature, which in turn increases the air temperature. The increased temperatures on the Mexican side promote an increase in the potential evapotranspiration

Figure 15. Sonoran desert vegetation discontinuity along the United States–Mexico border (photograph by R. Balling).

rates and further reduce the soil moisture levels. The end result of these processes is substantial warming in Mexico caused by overgrazing.

Because the overgrazing, resultant desertification, and landscape degradation occur over decades, it is reasonable to expect a relative warming trend for the areas that have experienced substantial desertification. To test this idea in the Sonoran Desert, I used the Jones et al. (1986a, 1986c) gridded temperatures to identify one point in Arizona where overgrazing had not occurred and another grid point in northern Mexico where overgrazing was severe (Balling, 1991a). When I subtracted the Mexican monthly temperature anomalies from the anomalies in Arizona, I found a 100-year differential warming of 0.32°C (0.58°F); the overgrazed Mexican landscape was warming at a statistically significant rate with respect to the area in Arizona.

I extended this same approach to other areas of the world where severe human-induced desertification was adjacent to areas of no desertification (Balling, 1991a). Fortunately, Howard Dregne (1977) of Texas Tech University had prepared a map for the United Nations showing where various classes of desertification had occurred. Based upon degree of landscape deterioration, Dregne defined four desertification categories. "Slight" desertification is marked by little deterioration of plant cover and soil, while the "moderate" class is characterized by plant cover deterioration, accelerated wind and water erosion, or an increase in soil salinity sufficient to reduce crop yields 10 to 50 percent. The areas of "severe" desertification have (1) undesirable forbs and shrubs in place of desirable grasses, (2) vegetation removal by sheet wind and water erosion, or (3) severe soil salinity and leaching problems; "very severe" desertification is an extreme condition of barren dunes, deep gullies, or salt crusts developed on nearly impermeable irrigated soils. Dregne's map shows approximately 30 percent of the land surface of the globe covered by these four desertification classes.

I used the 5° latitude by 10° longitude grid of Jones et al. (1986a, 1986c) to identify pairs of points around the globe where one grid point is within an area of severe desertification while an adjacent north or south land-surface grid point (exactly 552 km or 345 miles apart) is characterized by no desertification. Only 11 of these pairs so defined could be located in all (Table 3; Figure 16). A difference in temperature, ΔT, was determined for each month by subtracting the temperature anomaly for the grid point with no desertification from the temperature anomaly for the adjacent point with severe

Table 3. Desertification-Related Warming Trends.

Desert Region	Location of Pair Midpoint*	First Year	Months of Data	Warming Signal (°C/yr)	(°F/yr)
Sonoran	32.5°N 110°W	1881	1,176	0.0032	0.0058
Sahel	12.5°N 10°W	1896	950	.0063	.0113
Sahel	12.5°N 0°	1923	751	.0073	.0131
Sahel	12.5°N 5°E	1923	621	.0017	.0031
Sahel	12.5°N 10°E	1916	722	.0003	.0005
Sahel	12.5°N 15°E	1951	303	.0061	.0110
Sahel	12.5°N 20°E	1951	333	.0099	.0178
Sahel	7.5°N 25°E	1941	453	.0084	.0151
Turkestan	37.5°N 70°E	1893	1,113	.0044	.0079
Gobi	37.5°N 115°E	1922	597	.0072	.0130
Gobi	32.5°N 105°E	1932	586	.0001	.0002
Average		1921	691	.0050	.0090

* See Figure 16.

Figure 16.   Map of desertification categories and grid-point pairs (from Dregne, 1977).

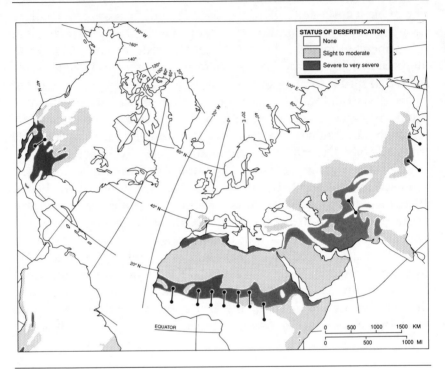

desertification. Using the new time series of ΔTs, I determined linear trend for each of the 11 pairs. The results (Table 3) indicate a warming trend associated with desertification at all of the 11 pairs. This warming signal in the data averages 0.005°C (0.009°F) per year; the results suggest a 100-year desertification-induced warming of 0.5°C (0.9°F) for these areas. The overgrazing in these arid and semiarid areas was producing a warming signal larger than the global warming observed over the past 100 years!

I argued that the data and methods used in these analyses would produce a substantial *underestimate* of the impact of desertification on the temperature record. The map prepared by Dregne is highly generalized, and was constructed with very limited information on the timing and degree of landscape degradation caused by human activities. In addition, the interpolation schemes used by Jones et al. (1986c) to generate the gridded temperature data from the widely scattered stations in the semiarid and arid regions of the world would tend to smooth spatial differences in the temperature trends. Despite these limitations, a significant desertification-related warming signal was identified at each of the 11 selected pairs. And given that 10 percent of the land area of the globe falls in the severe desertification category, while another 20 percent falls in the slight and moderate categories, the desertification warming signal in the global temperature record, like the urban heat island effect, accounts for between 0.01°C (0.02°F) and 0.10°C (0.18°F) of the global warming trend of the past century.

NON-GREENHOUSE EXPLANATIONS FOR RECENT TRENDS

We have undoubtedly seen an upward trend in the surface air temperature data generated by Jones et al. (1986c), but in addition to the greenhouse gases, many other forces could be causing the increase. We have already noted the influences of the reliability of the measurements, the existence of urban heat islands, and the effects of desertification. However, there remains the possibility, if not probability, that some of the variation and trend shown in Figure 13 could be explained by some non-greenhouse mechanism or external forcing of the climate system. For example, the temperature record of the globe may show the influence of volcanic eruptions, the appearance of large pools of warm or cold water over extensive ocean surfaces, or even variations in the output of the sun. We cannot forget that the climate system appears fully capable of producing variations and trends even in the absence of external influences. In this section, we will examine several of these external forcing mechanisms.

For years, many climatologists have believed that volcanic dust in the stratosphere can act to cool the earth (e.g., Baldwin et al., 1976; Pollack et al., 1976; Self et al., 1981; Sear et al., 1987; Bradley, 1988; Mass and Portman, 1989), particularly when the volcanic eruption emits large amounts of sulfur. The resultant dust and sulfuric acid particles in the stratosphere from the largest eruptions may increase the reflection of incoming radiation and ultimately act to cool the planet. Bradley (1988) examined Northern Hemisphere temperature records and showed that summer and autumn temperatures drop after large eruptions, but winter and spring temperatures appeared to be unaffected. Although some scientists question the volcanic link to global temperatures (e.g., Angell and Korshover, 1985; Ellsaesser, 1986), the work by Sear et al. (1987) provided rather convincing empirical evidence for the linkage. So the question remains: is some substantial amount of the variation or trend in the global temperatures of the past century related to dust in the stratosphere?

To answer this question, we need a long-term record of dust measurements in the stratosphere, and none exists. However, long-term sunshine records are available from a high-elevation (3,106 meters, 10,243 feet) station located in Sonnblick, Austria. These sunshine records are taken at a location where the transmittance of solar energy through the pristine atmosphere on selected days should be determined in a large part by stratospheric dust loads. Wu et al. (1990) used these sunshine records to generate estimates of stratospheric aerosol conditions for the globe. One problem with these data, as suggested by Jones and Wigley (1990a), is that the various time series supposedly estimating stratosphere dust levels are poorly correlated with one another. And although volcanic dust should disperse quickly and evenly throughout the stratosphere (Sedlacek et al., 1983), we cannot forget that the time series of Wu et al. (1990) is generated largely from the single location in Austria.

Despite the potential problems in using the estimates of stratospheric dust loads, the scatter diagram in Figure 17 shows a linkage between the dust loads and global temperatures. The figure shows that when the stratospheric dust levels are low, global annual temperatures tend to be high, and when the stratospheric dust levels are high, global temperatures appear to be lower. The equation describing the trend line takes the form:

$$T = 0.312 - 0.087 \text{ SAI} \qquad (3)$$

where SAI is the stratospheric aerosol index and T is the annual global temperature from the surface network. Over 20 percent of the variation in the global temperatures can be explained by stratospheric dust index using the linear trend line shown in Figure 17.

Equation 3 allows us to statistically control for the effects of dust in the stratosphere. For each year of record, the stratospheric aerosol index can be used to estimate the global temperature. When this estimate is subtracted from the actual global temperature value, a new temperature for each year is established. Note that this new global temperature has thus been statistically controlled for the amount of

Figure 17. Scatter diagram showing relation between Jones et al. (1986c) global temperature anomalies and the Wu et al. (1990) stratospheric aerosol index. Data are for the period 1895–1985.

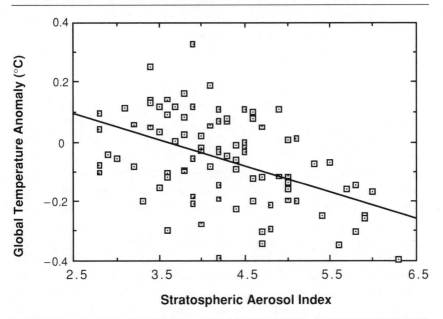

dust in the stratosphere (see Idso and Balling, 1991d). The 100-year rise in the temperatures that have been adjusted for stratospheric dust is 0.30°C (0.54°F); fully one-third of the original rise of 0.45°C (0.81°F) is eliminated, or accounted for, by the variations in the stratospheric dust. One-third of the global temperature trend of the past 100 years disappears when the stratospheric aerosol index is considered!

Along with research on the influence of volcanic dust on global climate, a number of climatologists have attempted to link variations in solar activity to the variations in global temperatures. Some groups argue in favor of such a linkage (e.g., Seitz et al., 1989; Reid, 1991), while others recently concluded that the effect of solar variability of global temperatures of the past century is very small (Hansen and Lacis, 1990; Kelly and Wigley, 1990; Wigley and Raper, 1990a). Similarly, the appearance of large warm pools or cold pools of water (called El Niño and La Niña) in the eastern portion of the tropical Pacific Ocean can exert some influence on global temperatures (e.g., Wright, 1985;

Bradley et al., 1987b; Wu et al., 1990). My own analyses of the relation between global temperature and the solar indices and various indices summarizing sea-surface conditions do not reveal any substantial explained variance in the global temperature records of the past century.

Table 4 summarizes some of the problems with accepting the 0.45°C (0.81°F) warming trend of the past century. While the table conveys a general sense of uncertainty about the established trend, the table shows that there are many candidates for reducing the trend. We see in the table that the statistical confidence interval around the trend may cause a 0.08°C (0.14°F) reduction, urban heat islands may cause a 0.05°C (0.09°F) reduction, overgrazing and desertification may lower the trend by a like amount, and stratospheric aerosol variations may lower the trend by 0.15°C (0.27°F). Due to the statistical interrelation among these contaminants and forcing mechanisms, it would not be correct simply to add these effects together and declare the remainder as the unexplained trend possibly linked to the greenhouse gases.

Again, it is important to recall that we do not know what the temperature of the earth would have been without any of the anthropogenic effects. Gordon (1991) recently concluded that the global temperatures of the past century have the appearance of a "random walk." That is, the variations in global temperatures (Figure 13) could be a result of purely random internal changes in the system—it is very possible to get these fluctuations and trends in the absence of any external forcing of the climate system. In fact, the climate models themselves predict substantial variations in the climate system that are not much different from the observed year-to-year fluctuations in the temperature record (see Houghton et al., 1991).

Table 4.   Summary of Adjustments to the 100-Year Global Warming Trend.

Global warming trend	+0.45°C
Potential adjustments:	
Reliability of the measurements	
Distribution of observation network	
Statistical uncertainty in trend	±0.08°C
Urban heat island effects	−0.01 to −0.10°C
Overgrazing/desertification	−0.01 to −0.10°C
Stratospheric aerosols	−0.15°C
Other climate forcings, internal variability	

Nonetheless, the amount of trend in global temperatures over the past century that is uniquely related to the observed 40 percent increase in equivalent CO_2 is likely to be less than the 0.45°C (0.81°F) established over the past 100 years. Too often we hear that the greenhouse gases are increasing and climatologists have found one-half degree Celsius of global warming in the past 100 years. As we have seen in this discussion, this warming could have occurred by chance, it may be the product of heat islands and measurement problems, some part of the warming may have been forced by a reduction in volcanic dust in the stratosphere, and very likely, some of the observed warming is directly related to the increase in greenhouse gases.

TIMING OF HISTORICAL TEMPERATURE INCREASES

An additional problem in linking the observed changes in surface temperatures to the buildup of greenhouse gases involves the timing of the warming. To illustrate this widely-recognized problem, two different figures have been prepared. The first (Figure 18) shows the 100-year annualized temperature values along with a smoothed version of the variations. Again, we see the warming from the beginning of the record to 1940, slight cooling from 1940 to the mid to late 1970s, and substantial warming from the late 1970s to the present. The plot indicates that in 1891, the smoothed temperature anomaly was −0.37°C (−0.67°F) while in 1990, the smoothed value was 0.32°C (0.58°F). The difference in smoothed values suggests an absolute warming of 0.69°C (1.24°F) over the past century; obviously, a few high or low values on either end of the 100 years can dramatically affect this estimate of warming. Nonetheless, the smoothed value for 1939 is 0.08°C (0.14°F), implying that 65 percent of the warming of the past century had occurred by the beginning of World War II. The bulk of the warming had occurred before the bulk of the trace gas increase!

A second way to examine this point is to calculate the linear warming from 1891 to a variety of years that followed. As an example, the warming between 1891 to 1990 was shown to be 0.45°C (0.81°F). That value, along with the confidence intervals associated with the calculation of the trend, is shown in Figure 19. However, the amount of warming between 1891 and 1940 is 0.34°C (0.61°F); these results also suggest that fully three-fourths of the linear warming of this century had occurred before the time of most rapid buildup of the greenhouse gases.

Figure 18.   Global temperature anomalies (based on 1950–1979 normals) for the period 1891–1990 along with smoothed (5-year averaged) values.

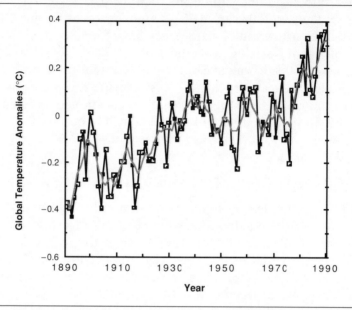

Figure 19.   Amount of global warming from 1891 to selected years in the Jones et al. (1986c) time series. Error bars are for the 0.95 level of confidence.

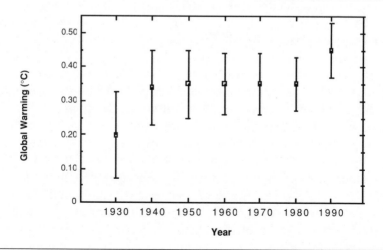

These analyses should not be used to underestimate the importance of the rather sudden warming that appeared in the record in the 1980s. Scientists promoting the "popular vision" of the greenhouse effect are quick to suggest that an inordinate number of years with record-setting warmth occurred very recently. In fact, the eight warmest years in the past 100 years are, in decreasing order, 1990, 1988, 1987, 1983, 1989, 1981, 1980, and 1986. Although the recent high temperatures can be explained, in part, by some of the contaminants to the record (e.g., heat islands, instrument changes, etc.) and to external forcing (e.g., low stratospheric dust levels, El Niño, solar output), the record-breaking warmth of the last decade is a signal in the surface-air temperature record that cannot be dismissed as unimportant. While some climatologists believe these recent years confirm the existence of an accelerated greenhouse warming (Jones et al., 1988; Thompson, 1989), others (e.g., Solow and Broadus, 1989) have noted that any upward trend necessarily means that the most recent years will be the warmest. Gordon (1991) noted that "random walk" variations tend to have their extreme values at the beginning and end of the record; the appearance of the extremes in the beginning and end of the climate record may have occurred by chance alone. Obviously, climatologists must conduct the research necessary to understand the underlying causes of the temperature patterns that appeared in the decade of the 1980s.

TROPOSPHERIC TEMPERATURE TRENDS

The analyses presented on global surface air temperature trends are potentially compromised by the many contaminants to the records. In an attempt to avoid these problems, researchers have identified data sets that are relatively free of heat-island effects, measurement errors, and other surface changes. One such data set comes from the upper-air soundings taken around the world; another comes from a satellite-based instrument package.

Each day at many locations around the globe, balloons ascend into the atmosphere carrying instruments capable of measuring temperature, pressure, winds, and moisture at many pressure levels. The instrument package relays the information back to a receiving station, and the details of upper-air conditions are determined. By the mid to late 1950s, these standard upper-air soundings were being taken in many parts of the globe.

Figure 20 shows the mean annual temperature from a 63-station, globally distributed radiosonde network for a layer in the atmosphere extending from 850 millibar (mb) level to the 300 mb level (the sea-level atmospheric pressure is near 1,000 mb and the top of the atmosphere would be near 0 mb). Generally, this 850–300 mb layer extends from about 1.4 km (0.9 mile) to 9.1 km (5.7 miles) above sea level. Many of the numerical models of climate suggest that this layer of the atmosphere should warm even more than the surface as greenhouse gases are increased. Some analyses (e.g., Hansen et al., 1984; Schlesinger and Mitchell, 1987; Mitchell et al., 1990) suggest that the warming in the 850–300 mb level should be 50 percent more than what is observed at the surface, particularly in the tropical areas.

Figure 20.   Time series of global 850–300 mb temperature anomalies (Angell, 1988) and global surface air temperatures (Jones et al., 1986c). Data are available in Boden et al. (1990).

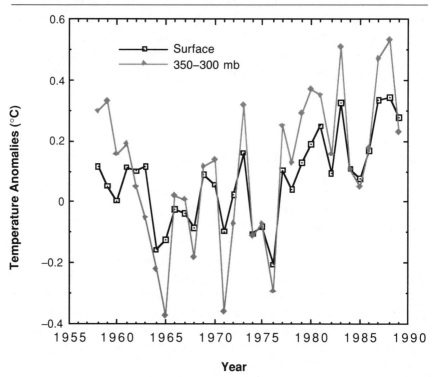

And in fact, over the 1958–1989 time period, the 850–300 mb temperature anomalies (Angell, 1988) show warming of 0.098°C (0.178°F) per decade (Figure 20). Over the same period, surface air temperatures from Jones et al. (1986c) show a 0.081°C (0.145°F) warming per decade. Obviously, warming has been observed in the mid-troposphere and, as the models predict, the warming in the 850–300 mb layer is greater than the warming observed at the surface. Unfortunately, the upper-air soundings do not extend back in time to allow analyses on a 100-year time span. However, the analyses of the upper-air temperature data do not alter the basic conclusions from the examination of the surface air temperature measurements.

The recent satellite-based measurements of mid-tropospheric temperatures (Spencer and Christy, 1990; Spencer et al., 1990; Diaz, 1990) provide another measurement of global temperature that must be considered in our discussion. A passive microwave sensor in space can fly a polar orbit and have an opportunity to make measurements over the entire globe. Measurements made by the 53.74 GHz channel of the microwave sounding unit would be sensitive to the thermal emission of molecular oxygen in the middle troposphere. The measurement would not be particularly affected by changes in water vapor, cloud variations, or changes at the surface, nor should temperature trends in the overlying stratosphere strongly influence the microwave data (Gary and Keihm, 1991). The data would cover the entire planet, and the measurement should be accurate to within a few hundredths of a degree.

Figure 21 shows the passive microwave estimate of annual global temperature over the period 1979–1989 along with the surface data of Jones et al. (1986c) and the radiosonde data of Angell (1988). The patterns depicted by these curves are very consistent among the data sets, leading one to conclude that the long-term patterns for the surface data are largely representative of the entire troposphere. However, a different conclusion may be reached in examining Figure 22, which displays the rate of temperature increase for the surface, radiosonde, and satellite estimates. Here we see that the rate of change in temperature over the 1979–1989 period as determined by the surface network is nearly 20 times larger than the value established for the satellite passive microwave system, and over 5 times larger than the value for the radiosonde network. The low number of years represented by these different temperature data does not allow us to conclude that these differences are statistically significant; however, these differences will likely become significant as more years of satellite data are collected.

Figure 21. Time series of satellite-based global temperature anomalies (Spencer and Christy, 1990), global 850–300 mb temperature anomalies (Angell, 1988), and global surface air temperatures (Jones et al., 1986c) over the period 1979–1989.

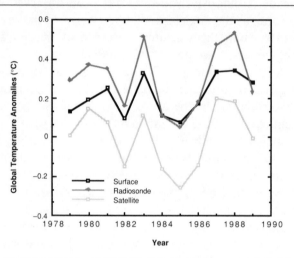

Figure 22. Annual global warming rate over the period 1979–1989 as measured by the radiosonde network for the 850–300 mb level (Angell, 1988), the satellite microwave data (Spencer and Christy, 1990), and the surface (land and sea) air temperature measurements (Jones et al., 1986c).

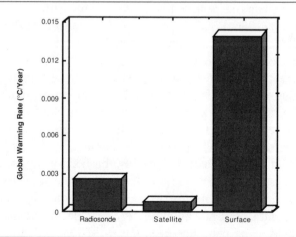

STRATOSPHERIC TEMPERATURE TRENDS

The stratosphere generally extends from approximately 10 km (6.2 miles) to about 50 km (31.2 miles) in height. In the lower part of the atmosphere discussed so far, the temperatures generally decrease with an increase in height. However, in the stratosphere, the temperature either varies little with height or increases with height; therefore, the stratosphere is considered to be quite separate from the underlying troposphere, and climatologists tend to treat these layers separately in their analyses of future climate change.

Virtually every climate model has predicted that stratospheric temperatures will decrease as the greenhouse gases build up in the atmosphere. In many models, the decrease in temperatures in the stratosphere is predicted to be as large as the increase in tropospheric temperatures (e.g., Mitchell et al., 1990). Although not a major part of the "popular vision" of the greenhouse effect, cooling in the stratosphere is one of the most consistent predictions of the numerical climate models. Fortunately, we have limited temperature data from the stratosphere that may be used to test for cooling that may be occurring already.

Angell (1988) provided radiosonde data across the 63-station global network for the 100 to 30 mb level. This level is within the stratosphere and is, on average, 16 and 24 km (10 and 15 miles) in height. A plot of these 100–30 mb temperatures from the period 1958 to 1989 reveals a statistically significant decrease of 0.33°C (0.59°F) per decade (Figure 23). The very cool stratospheric temperatures recorded in the 1985 to 1989 period are related to substantial cooling seen in the south polar region. But despite uncertainties in the measurement of stratospheric temperature, we can conclude that the recent temperature trends in the 100–30 mb layer of the atmosphere are consistent in sign and magnitude with the predictions of the models for a doubling of equivalent CO_2.

OTHER OBSERVED GLOBAL-SCALE CLIMATE CHANGES

So far in this chapter, we have concentrated our attention on changes taking place in atmospheric temperatures near the surface, in the middle of the troposphere, and in the stratosphere. Because the greenhouse effect is likely to produce changes in many other climatic variables,

Figure 23. **Annual global stratospheric (100–30 mb) temperatures over the period 1958–1989 as measured by the radiosonde network (Angell, 1988). Data are available in Boden et al. (1990).**

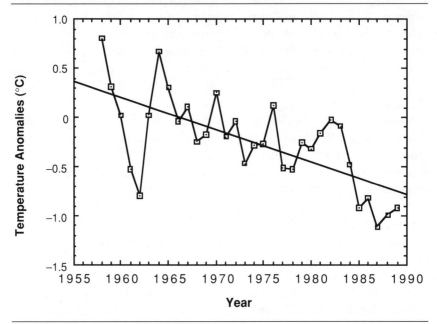

it may be useful to examine changes that have been observed in sea levels, global sea-surface temperatures, cloud cover, precipitation, and snow and ice cover. The following points summarize the changes observed in these variables.

The concept of a rising sea level is another of the many frightening components in the "popular vision" of the greenhouse effect. Initially, some popular reports had sea levels rising by more than 10 m (more than 30 feet) inundating many coastal cities. However, as with many other components in the greenhouse debate, the estimates of sea level rise have been falling (Jones and Henderson-Sellers, 1990), and many uncertainties surround any estimates of past or future trends.

The identification of a signal in the sea level is probably even more complicated than the search for temperature or precipitation signals in the global record. The usual problems exist in collecting sea-level data: tide gauge records are often too short, the instruments for making the measurements have changed through time, and too many of the records come from Northern Hemisphere mid-latitudes while too few records exist for the rest of the world. Further complicating the issue is the fact that the sea level at a location is affected by meteorological conditions. Changes in winds and atmospheric pressure at a given location can directly change the sea-level measurement; a shift in an ocean current also can influence the sea-level measurement. But most important, the sea-level values are influenced by up and down movements of the land areas. The land surfaces move vertically with respect to sea level due to any number of influences involving extraction of groundwater or oil, sedimentation, or tectonic events such as post glacial rebound. All of these factors must be accounted for before any reliable trends in an already imperfect sea-level record can be identified.

Recognizing the problems in the data, but also recognizing the importance of determining trends in sea level, many scientists have been drawn to the problem (excellent reviews are available in Warrick and Oerlemans, 1990; Warrick and Wigley, 1990; Emery and Aubrey, 1991). Within the past 100 years, the evidence is that global sea level has risen at a rate of approximately 1.1 mm/yr (0.04 inch/yr), yielding a total rise of approximately 112 mm (4.4 inches) over the entire time period. Thermal expansion of the oceans, caused by the warming of the planet, has probably caused about 40 percent of this increase while the melting of glaciers and small ice caps has caused another 40 percent of the increase in sea levels. Melting of ice and snow in Greenland most likely accounts for the remaining portion of sea level rise.

The future rise will obviously depend on the amount of warming experienced in the coming decades. However, given the observed 112 mm rise during a period when global temperatures rose about 0.4°C (0.7°F), estimates of sea level rise from 80 mm (3.1 inches) to 290 mm (11.4 inches) from now until the time of CO_2 doubling are probably very reasonable (Warrick and Oerlemans, 1990). The projected rise in sea level is far less in magnitude, and far less threatening, than the enormous rise projected as a part of the "popular vision." Very simply, if we do not get substantial global warming, we will not realize any catastrophic rise in sea levels.

The pattern in global sea-surface temperatures over the past 100 years is similar to the patterns described for surface air temperatures (Bottomley et al., 1990; Folland et al., 1990). Given that over 70 percent of the planet is covered by the oceans, and given the coupling between the ocean and atmosphere, the similarity in the two patterns is expected.

In many respects, the measurement of precipitation is more difficult and prone to error than the measurement of temperature. Not only is precipitation more variable in time and space than temperature, but the actual measurement of precipitation is hindered by wind, evaporation of collected rainfall, and blowing snow. Despite these problems, and recognizing that the data are not readily available for ocean areas, there is still evidence from much of the globe that precipitation has increased over the past century. For example, the annual precipitation index for all land areas of the globe is presented in Figure 24 for the period 1891 to 1986 (Diaz et al., 1989). This index represents the mean of percentiles for grid points around the globe; values above

Figure 24.   Annual precipitation index for land surfaces of the globe for the period 1891–1986. Values are means of percentiles for grid points around the globe (e.g., 0.50 is an average precipitation value, 0.45 is below average, and 0.55 is above average). Data are from Diaz et al. (1989).

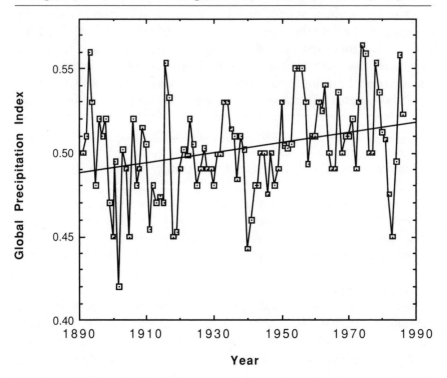

0.50 indicate wet years and values below 0.50 are indicative of dry years. The plot shows a statistically significant increase in the precipitation index over the past 100 years of approximately 6 percent. This value is in agreement with Vinnikov et al. (1990), who also found a 6 percent increase in precipitation over the past 100 years for the land-surface band between 35°N and 70°N.

Given the observed increase in precipitation, one would expect an increase in cloudiness over the past century, and in fact, such an increase has been observed over many land areas and over the oceans. Results by Henderson-Sellers (1986a, 1986b, 1989) and McGuffie and Henderson-Sellers (1988) suggest that the cloudiness has increased between 5 and 10 percent over the past century; data presented by Warren et al. (1988) are consistent with a total cloud cover increase over the oceans during the past 50 years.

Finally, one may expect that the slight warming observed over the past century should have caused a reduction in glacial extent, sea ice coverage, and snow-covered areas over land. However, the increase in precipitation observed for much of the planet may act to increase the extent of glaciers and the coverage by snow. The issue is particularly difficult to resolve given the uncertain and short-term measurements of glaciers, sea ice, and snow cover. Nonetheless, Robinson and Dewey (1990) concluded from their analyses of satellite photographs that the snow coverage had decreased in the warm decade of the 1980s. Folland et al. (1990) reviewed a number of studies on sea ice extent and concluded that no significant long-term trends could be identified in the various data sets. The same is somewhat true for the glaciers of the world. Some studies (e.g., Grove, 1988; Patzelt, 1989) suggest a general recession in glaciers over the past century while others (e.g., Wood, 1988b) suggest that the number of retreating glaciers is declining while the number of advancing glaciers is increasing. The glacial response to the record-breaking warmth in global temperatures is simply not known at this time.

OBSERVED VERSUS PREDICTED GLOBAL CLIMATE CHANGE: A SUMMARY

Over the past 100 years, equivalent CO_2 levels have increased by nearly 40 percent, rising from slightly less than 310 ppm to over 430 ppm. Given this increase in equivalent CO_2, climate models typically predict a warming of at least 1.0°C (1.8°F), a small reduction in the diurnal temperature range, a 2 to 5 percent increase in precipitation, some increase in cloud cover (depending on how the models calculate the cloudiness), a general decrease in the coverage of snow, glaciers, and sea ice, and a moderate rise in sea level. If one is going to accept the predictions of the models for a doubling of equivalent CO_2, the models should be able to replicate the observed climate changes over the past 100 years when a significant increase in equivalent CO_2 occurred.

So how did the models score? In many respects, the models did remarkably well simulating the observed changes to the global climate system. The observed increases in cloud coverage and precipitation are generally consistent with model expectations given a 40 percent increase in equivalent CO_2; in fact, the models have underestimated the change in these moisture-related climate components. Although not conclusive, the data from glaciers, snow coverage, and sea ice extent certainly do not contradict the models. A rise in temperature was observed, but the increase appeared to be far less than the model predictions. Not surprisingly given the smaller than expected increase in temperature, the rise in sea level has been less severe than what would have been predicted from the various models. In addition, some evidence suggests that the diurnal temperature range was decreasing faster than the predictions by some of the models.

From this discussion alone, a new vision of the greenhouse world begins to emerge. This vision is consistent with the general predictions of the models and it is consistent with the observations of climate change over the past century. When compared to the climate of the past 100 years, the greenhouse world (at 600 ppm) is likely to receive more precipitation; an increase in precipitation near 10 percent may be reasonable. Cloud coverage is likely to increase about the same amount. Global temperatures will rise, but possibly by no more than 1.0°C (1.8°F); the decrease in diurnal temperature range would suggest that the warming will largely occur at night. Because the afternoon temperatures do not rise significantly, evaporation rates do not increase (as seen by the "popular vision"), and soil moisture levels may increase. The moderate rise in global temperature would likely translate into only moderate rises in sea level. These global patterns are internally consistent with one another, they are consistent with climate changes observed during a time when equivalent CO_2 increased by

40 percent, and they are consistent generally with the predictions of the numerical models. We are already watching the greenhouse effect unfold before our eyes, and the observational evidence is not pointing to disaster.

5

OBSERVED HEMISPHERIC AND LATITUDINAL CLIMATE CHANGE

In the previous chapter, we examined historical climate changes that had been observed over the past 100 years for the planet as a whole. In many cases, we used one number to represent the state of some climate variable (e.g., temperature, precipitation) for a given year and for the entire globe. We found that the earth was experiencing significant increases in both temperature and precipitation over the past 100 years. And while the recent policy statement made by the American Meteorological Society (1991, p. 58) on global climate change states that "today's climate models provide little or no useful and consistent information on regional distributions of climate change," we may gain more insights into how the climate has responded to a 40 percent increase in equivalent CO_2 by examining the hemispheric and latitudinal distributions of climate change. As with other analyses presented thus far, some of these hemispheric patterns support while others pose serious challenges to the climate model predictions.

HEMISPHERIC TEMPERATURE TRENDS

The Jones et al. (1986a, 1986b, 1986c) gridded temperature data base provides every opportunity to calculate the trends in temperature for the Northern and Southern Hemispheres. In fact, the global trend is rarely presented in the professional literature without the trends for the two hemispheres. As we examine these patterns, it is important to remember that (1) the Northern Hemisphere has a more complete sampling network for producing the temperature measurements, (2) the oceans comprise 61 percent of the Northern Hemisphere and 81 percent of the Southern Hemisphere, and (3) due to this difference in ocean coverage, the models generally predict a larger and more immediate increase in temperature in the Northern Hemisphere when compared to the Southern Hemisphere (Stouffer et al., 1989).

A plot of the Northern Hemisphere temperature curve from 1891 to 1990 reveals a pattern of basically no trend in temperatures from 1891 to about 1920, a sharp rise in temperature from 1920 to 1940, cooling from 1940 to the late 1970s, and record warmth in the 1980s (Figure 25). Over the entire 100-year period, a trend line shows that

Figure 25. Northern Hemisphere temperature anomalies (based on 1950–1979 normals) for the period 1891–1990. Data preparation is described by Jones et al. (1986a); actual data are available in Boden et al. (1990).

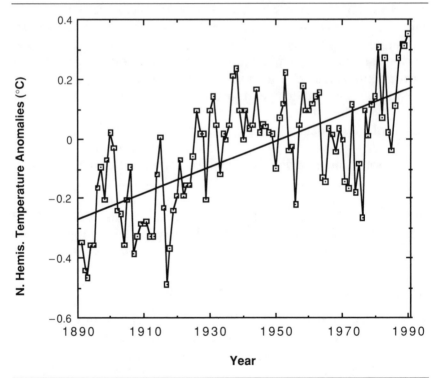

the amount of warming in the Northern Hemisphere was 0.44°C (0.79°F). However, as seen in Figure 26, the amount of linear warming between 1891 and 1940 is 0.41°C (0.67°F), the warming between 1891 and 1950 is 0.44°C (0.79°F), and the warming between 1891 and 1960 is 0.45°C (0.81°F). Once again, we see that a great deal of the warming of the past century had already occurred before the largest buildup in the greenhouse gases.

The pattern for the Southern Hemisphere is somewhat different from that observed in the Northern Hemisphere. Figure 27 shows that the surface air temperatures in the Southern Hemisphere generally increased from 1891 to the mid-1940s. Near the end of World War II, the temperatures dropped sharply, and since that time the Southern Hemisphere has shown a rather steady increase in temperature. Again, the 1980s are dominated by record-setting warmth. Figure 28 shows that the amount of warming from 1891 to 1990 was 0.47°C (0.85°F)

Figure 26. Amount of Northern Hemisphere warming from 1891 to selected years in the Jones et al. (1986a) time series. Error bars are for the 0.95 level of confidence.

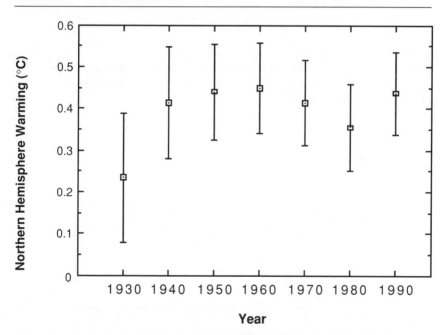

Figure 27.   Southern Hemispheric temperature anomalies (based on 1950–1979 normals) for the period 1891–1990. Data preparation is described by Jones et al. (1986b); actual data are available in Boden et al. (1990).

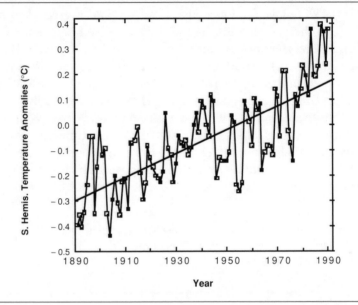

Figure 28.   Amount of Southern Hemisphere warming from 1891 to selected years in the Jones et al. (1986b) time series. Error bars are for the 0.95 level of confidence.

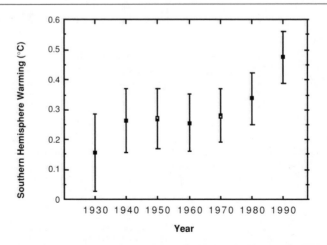

while from 1891 to 1940, the warming in the Southern Hemisphere was only 0.26°C (0.47°F).

Several substantial differences may be noted in the temperature pattern for the two hemispheres. First, the Southern Hemisphere warmed more than the Northern Hemisphere. Although the difference between them over the past 100 years may appear to be small, only 0.04°C (0.07°F), the difference is statistically significant. In addition, the 850–300 mb temperatures from Angell (1988) show a 0.05°C (0.09°F) warming in the Northern Hemisphere over the period 1958–1989 and a 0.56°C (1.01°F) warming in the Southern Hemisphere over the same period. Despite model predictions to the contrary, the Southern Hemisphere, with its large expanse of oceans, is warming significantly faster than the Northern Hemisphere! Second, the warming in the Southern Hemisphere has been much more steady than the warming of the Northern Hemisphere. Over 90 percent of the Northern Hemisphere warming for the past 100 years had already occurred by 1940, when only 55 percent of the total Southern Hemisphere warming had occurred. These and other differences between the hemispheric temperature patterns remain as important parts of the ongoing greenhouse debate (see Karoly, 1987).

A plot of the differences in the hemispheric temperatures (Northern minus Southern) reveals an interesting pattern of the Northern Hemisphere warming with respect to the Southern Hemisphere up to the late 1950s (Figure 29). Rather abruptly, the Northern Hemisphere began to cool with respect to the Southern Hemisphere (or conversely, the Southern Hemisphere warmed with respect to the Northern Hemisphere) from the late 1950s to the end of the record. The surface observations of Jones et al. (1986a, 1986b) suggest that between 1958 and 1989, the difference in hemispheric temperature trends was 0.19°C (0.34°F). Something was either retarding warming in the Northern Hemisphere or stimulating warming in the Southern Hemisphere. The 850–300 mb tropospheric temperatures of Angell (1988) show a large difference of 0.51°C (0.92°F) in the 1958–1989 hemispheric trends; once again, the evidence suggests an accelerated warming in the Southern Hemisphere and/or a decelerated warming in the Northern Hemisphere.

The reason for the hemispheric differences is not easy to reconcile— the greenhouse gases are very evenly distributed around the planet, and several contaminants to the record (heat islands, desertification) should be producing differential warming in the Northern Hemisphere, not the Southern Hemisphere. Any comprehensive vision of climate

Figure 29.   Differences in hemispheric temperature (Northern minus Southern) over the period 1891–1990. Hemispheric data are described by Jones et al. (1986a, 1986b) and are available in Boden et al. (1990).

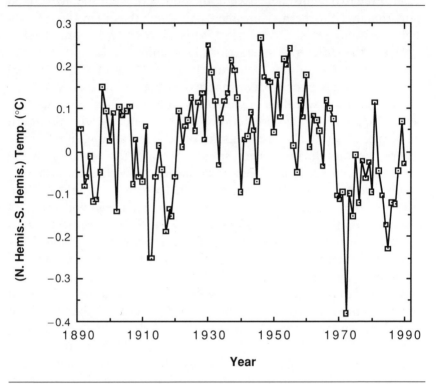

change into the next century must be able to account for this hemispheric difference in temperature trend. As we shall see in Chapter 7, the role of other pollutants may provide a plausible explanation.

LATITUDINAL TEMPERATURE TRENDS

The analyses of the previous section show that over the past 100 years, both hemispheres have warmed, the Southern Hemisphere more than the Northern Hemisphere. The models certainly predict warming in both hemispheres, but they fail to account for the differential warming rates. In this section, we will examine the changes in annual temperatures for three latitudinal bands in the Northern Hemisphere.

These analyses would be restricted in the Southern Hemisphere by the scarcity of long-term temperature data within the selected latitudinal bands.

The first area is the Arctic region, defined as the area extending from the Arctic Circle (66.5°N) to the North Pole (90°N). Many of the climate models predict that as equivalent CO_2 levels are increased, this polar area (covering approximately 8 percent of the Northern Hemisphere) will experience double the warming of the entire planet (e.g., Mitchell et al., 1990). In other words, models predicting a 4.0°C (7.2°F) response of global temperature to a doubling of CO_2 suggest that the Arctic region could warm by at least 8.0°C (14.4°F). A plot of the temperature anomalies for this region (Figure 30) shows essentially no trend from 1891 to 1920, rapid warming from 1920 to 1940, cooling from 1940 to the mid-1960s, and a slight warming since that time. The range of

---

Figure 30.   Temperature anomalies for the Arctic region (66.5°N–90°N) over the period 1891–1988; data are described by Jones et al. (1986a.)

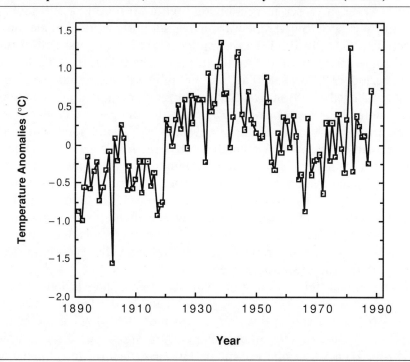

Year

---

annual temperatures experienced in the Arctic area is approximately 3.0°C (5.4°F); this range is very much larger than the range seen for the globe or the Northern Hemisphere. Along with this temporal variability, the region has shown great spatial variability as well. Generally speaking, warming has occurred over Alaska and northern Asia while cooling has occurred over northern Europe, the northern Pacific, and Greenland (Hansen and Lebedeff, 1987; Jones, 1988; Trenberth, 1990).

Within this pattern of wide temporal and spatial variability, a linear warming of 0.61°C (1.10°F) exists over the Arctic for the period 1891 to 1988. Some scientists (e.g., Lachenbruch and Marshall, 1986; Wadhams, 1990; Gloersen and Campbell, 1991; Quadfasel et al., 1991) have found evidence of thinning of Arctic ice and linked these changes in snow and ice to the general increase in the Arctic temperatures. Because this amount of warming is 44 percent greater than what is observed for the planet as a whole over the same time period, one may conclude that warming in the Arctic is broadly consistent with the predictions of the models.

However, a closer inspection of Figure 30 reveals a substantial problem in accepting Arctic warming as supportive of the model projections. Virtually all of the warming occurred in the first half of the record; temperatures in the Arctic area *cooled* by 0.42°C (0.76°F) during the period 1940 to 1988. The 850–300 mb tropospheric temperature data of Angell (1988) suggest that the surface air temperature data may have actually *underestimated* the amount of cooling in the 1958 to 1988 time period. Although Kalkstein et al. (1990) found some warming in selected air masses in Alaska, the bulk of the Arctic region has shown an overall statistically significant cooling during the time of greatest increase in equivalent CO_2 levels! The ocean thermal lag and the time required to melt ice and snow, thereby exposing the underlying terrestrial surface and promoting warming, are often offered as plausible explanations for no warming in the Arctic. Nonetheless, the temperature patterns in the Arctic region do not provide much empirical support for the greenhouse predictions.

The mid-latitude region of the Northern Hemisphere is defined here on the north by the Arctic Circle (66.5°N) and by the Tropic of Cancer (23.5°N) on the south; this area covers more than half of the hemisphere. A plot of the annual temperature anomalies for this area (Figure 31) is very similar to the pattern described for the global and

Figure 31.   Temperature anomalies for the Northern Hemisphere mid-latitude region (23.5°N–66.5°N) over the period 1891–1988; data are described by Jones et al. (1986a).

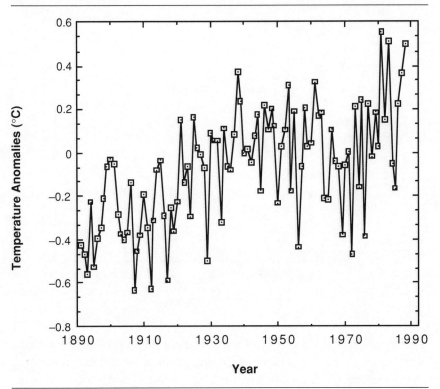

Northern Hemisphere records. Warming dominated from 1891 to 1940, cooling occurred from 1940 to the mid-1970s, and the 1980s were characterized by record warmth. This mid-latitude region experienced a linear temperature increase of 0.53°C (0.95°F) over the 1891 to 1988 period of record, which is 25 percent higher than the warming experienced by the globe and 31 percent higher than the warming of the Northern Hemisphere. The models predict that this area should warm more than the hemisphere or the globe; however, given a 40 percent increase in the equivalent CO_2 levels, the amount of warming is far less than what the models would predict for the past 100 years.

The mid-latitude zone provides an opportunity to test the seasonality of temperature trends over the 1891 to 1988 period. Figure 32 displays

Figure 32. Observed linear warming and confidence intervals by month for the Northern Hemisphere mid-latitude region (23.5°N–66.5°N) over the period 1891–1988; data are described by Jones et al. (1986a).

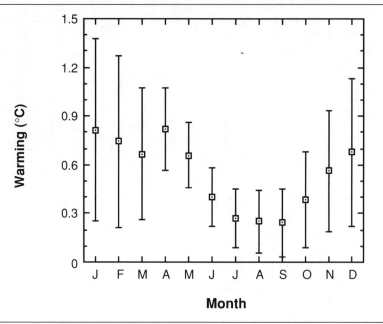

the warming trend, along with the confidence intervals, for each calendar month. As can be seen in the figure, the months of greatest warming are December–May and the months of least warming are June–November. More specifically, the average amount of warming in December–May is 0.73°C (1.31°F) while the warming in June–November averages 0.35°C (0.63°F): twice as much warming occurs in the December–May period when compared to the June–November period. The models tend to predict greatest warming in late autumn and winter and least warming in the summer season (e.g., Hansen et al., 1984; Schlesinger and Mitchell, 1987; Mitchell et al., 1990). Therefore, the seasonal pattern determined for the mid-latitudes of the Northern Hemisphere is broadly consistent with the predictions of the models.

The tropical region, bounded by the Tropic of Cancer (23.5°N) and the Equator (0°) and encompassing about 40 percent of the hemispheric surface area, shows a pattern of warming from 1891 to 1940, some cooling from 1940 to the mid-1970s, and moderate warming from the mid-1970s to the late 1980s (Figure 33). Noticeably missing are the many record-breaking years in the 1980s found in the hemispheric and global temperature patterns. The linear warming over the 1891 to 1988 time period is 0.24°C (0.43°F), which is well below the amount of warming the models would predict for a 40 percent increase in equivalent CO_2.

Patterns in the Southern Hemisphere are likely to be similar to those found in the Northern Hemisphere. The tropical and mid-latitude areas cover more than 90 percent of the surface area of the Southern Hemisphere, and by definition they must have a temperature pattern

Figure 33.   Temperature anomalies for the Northern Hemisphere tropical region (0°–23.5°N) over the period 1891–1988; data are described by Jones et al. (1986a).

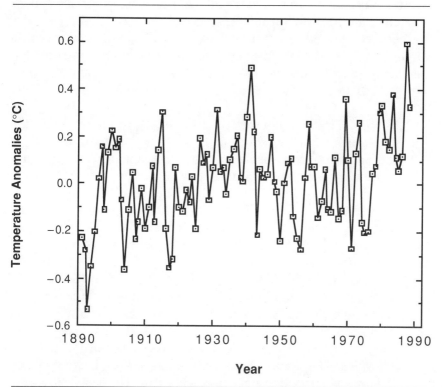

similar to the pattern for the entire hemisphere. Although long-term temperature records are obviously sparse in Antarctica, researchers (e.g., Jones et al., 1986b) have found Antarctica to be decoupled from the climate patterns of the remainder of the Southern Hemisphere. Some scientists have found cooling in Antarctica, others have found little evidence for any trend, and still others have noted some warming in the region (e.g., Mo and van Loon, 1985; Jones et al., 1986b; Jones, 1988; Sansom, 1989; Trenberth and Olson, 1989; Jones, 1990; Doake and Vaughan, 1991). While the surface air temperature data in Antarctica are inconclusive in determining the existence of trends, the Angell (1988) 850–300 mb tropospheric temperature data for the area, despite their shortcomings, show a warming of 0.77°C (1.39°F) over the 1958 to 1989 time period.

HEMISPHERIC AND LATITUDINAL PRECIPITATION TRENDS

As we have seen, several large global-scale data sets have been assembled allowing careful analyses of temperature trends for the globe, the hemispheres, and various latitudinal bands. In a similar way, several large-scale precipitation data sets for land areas have been generated and analyzed by Bradley et al. (1987a), Diaz et al. (1989), and Vinnikov et al. (1990). General findings of these analyses include the following:

1. As noted earlier and presented in Figure 24, precipitation for the land areas of the globe has increased over the past 100 years.
2. The Southern Hemisphere has shown a statistically significant increase in precipitation since 1891 while the Northern Hemisphere has shown no significant trend. The Southern Hemisphere precipitation levels increased from 1940 to the present while the Northern Hemisphere levels have tended to decrease since 1950.
3. Within the Southern Hemisphere, the increase in precipitation has occurred within all latitudinal bands and within all seasons. The greatest increases have occurred in the spring (September–November) and autumn (March–May) seasons; the summer (December–February) period has the smallest increase.
4. Precipitation changes in the Northern Hemisphere are far more complex than the trends in the Southern Hemisphere. On a hemispheric basis, the winter season has shown a rather steady decline in precipitation since the early 1950s. The decline in spring

season precipitation began in the late 1970s, while the summer and autumn seasons show a pattern of abrupt reductions in precipitation in the early 1980s. Somewhat surprisingly, the changes in precipitation through time have varied tremendously from one latitudinal band to the next. Virtually all of the recent reduction in precipitation has occurred in the tropical region, particularly in the African Sahel. This reduction in precipitation is particularly strange given the increase in precipitation found in the Southern Hemisphere tropics. While the Northern Hemisphere tropical region shows a decline in precipitation, the mid-latitudes have shown a rather steady upward trend since the mid-1940s.

5. The Southeast Asia monsoon has shown a slight decline in precipitation in the most recent 25 years, but overall, no significant trend up or down is found in the monsoon precipitation totals.

With respect to the greenhouse issue, we must ask whether these observed changes in precipitation are consistent with the predictions of the numerical models. Diaz et al. (1989, p. 1195) stated, "The observations are only broadly consistent with zonally averaged profiles of precipitation changes derived from general circulation model (GCM) simulations of climate using doubled atmospheric CO_2 concentrations, although we note that there is considerable variability in precipitation response from one model to another." The reduction in precipitation in the Northern Hemisphere tropical zone is really the problem; otherwise, the observed increase in precipitation over the rest of the planet is generally consistent with the model predictions.

CHAPTER SUMMARY

At the global scale, we found the models were predicting increasing temperature and precipitation as the greenhouse gases increase in concentration. Given the 40 percent increase in equivalent CO_2 over the past 100 years, we saw that the globe had warmed slightly and become wetter (at least over the land areas). At a highly generalized level, the model predictions are consistent in sign, but not in magnitude, with the global climate records of the past century. However, as we move from the global to the hemispheric scale, the predictions of the models find even less empirical support.

As we saw in this chapter, six fundamental points emerge in the analysis of hemispheric and latitudinal climate trends. First, the amount of warming experienced over the past 100 years in either hemisphere is considerably less than expected given the observed increase in equivalent CO_2 levels. Second, despite model predictions to the contrary, the Southern Hemisphere has warmed more than the Northern Hemisphere, particularly in the most recent half-century. Third, some evidence shows that the long-term warming is greatest in the polar regions and least in the tropical regions; this distribution is broadly consistent with the predictions of the models. Fourth, the models predict greatest warming in the low-sun season and the least warming in the high-sun season, and generally this is observed in the mid-latitude temperature records. Fifth, the Arctic region as a whole, where the greatest amount of warming should be observed, has shown cooling over the past 50 years. And sixth, the observed patterns for precipitation trends are only broadly consistent with the highly varied predictions of the climate models.

Therefore, we again see evidence that a doubling of CO_2 will produce a temperature response on the low end of the predictions, possibly near 1.0°C (1.8°F). The warming will likely occur more at night than during the day, the warming will probably occur most in the high-latitude locations, and the warming will be revealed largely during the winter season; cloudiness and precipitation will increase in most locations. This internally consistent change in the climate is supported by model predictions, analyses of the global climate records, and now by the analyses of the hemispheric and latitudinal records.

6

GREENHOUSE EFFECTS IN THE UNITED STATES?

Americans have been told that their temperatures will increase, droughts will plague the agricultural heartland, extreme high temperatures will occur much more frequently, hurricane and severe storm activity will increase, streams and rivers will dry up, sea level will rise and inundate low-lying coastal areas, wildfires will become more common, and on and on. And the culprit for this apocalypse is, as always, the buildup of the greenhouse gases. However, the climate patterns in the United States observed over the past century have not been particularly supportive of this apocalyptic view.

In this chapter we will explore, in some detail, the kinds of changes that have been observed within the conterminous United States over the past century. As we have done throughout the book, the observed changes in a variety of climate variables will be compared to the model predictions for the United States area. From the outset, it is critical to realize that the area of the conterminous United States (excluding the U.S. portion of the Great Lakes) covers only 1.53 percent of the surface of the globe; the conterminous United States accounts for slightly more than 5 percent of the total landmass of the planet. What happens to the climate of the United States may seem vitally important within the country, but in terms of global climate change, what happens in the continental United States has little effect on the planetary trends.

Nonetheless, a substantial portion of the scientific literature on the greenhouse effect has focused on potential changes in the climate patterns of the United States. Given the number of climatologists in the country, the number of professional journals published within its borders, and the number of decisionmakers concerned about climate changes in the United States, the importance of climate trends in this 1.53 percent of the planet has been inflated with respect to the greenhouse debate. However, the "popular vision" is full of predictions that are highly specified for this area.

PATTERNS IN U.S. TEMPERATURES

One of the many reasons scientific investigators are drawn to the analysis of climate change in the United States is the availability of long-term, reasonably homogeneous weather and climate records. One outstanding data base is the United States Historical Climatology Network (HCN) developed by climatologists at the National Climatic Data Center in Asheville, North Carolina (Quinlan et al., 1987). The HCN was developed to detect regional changes in climate as opposed to station-specific changes that may have occurred over the past century. The network consists of 1,219 stations rather evenly distributed through the conterminous United States. To be included in the HCN, each station had to be active in 1984, have at least 80 years of temperature and precipitation data, have few missing data, and have experienced few station relocations over the collection period. Every effort was made to avoid major metropolitan areas where heat island effects could dominate the record; the median population for towns in which the stations were located was only 5,832 in 1980 (Balling and Idso, 1989).

An extremely complex set of quality control measures was applied to the temperature and precipitation records from the 1,219 stations. The history of each station was carefully examined, and the records were statistically adjusted for changes in the time of observation, the actual instruments used in the measurements, the location of the instruments, and the location of the station. Homogeneity tests were run on the temperature and precipitation data, and ultimately a time series of mean monthly temperatures (including mean maximum, mean minimum, and average temperature) and monthly precipitation totals was established for the past century. Recognizing that climate data from small towns could be affected to some degree by growing urban heat islands (Balling and Idso, 1989), the temperature data were further adjusted for any growth in the population (Karl et al., 1988). Due to a substantial drop in the number of reporting stations prior to 1900, all analyses of the HCN urban-adjusted data presented in this chapter will extend from 1901 to 1987.

Figure 34 presents the mean annual temperatures for the continental United States during the twentieth century. Much like the temperatures for the Northern Hemisphere and for the globe, the U.S. temperatures show no trend from 1901 to 1920, warming from the early 1920s to the late 1930s, cooling through the 1970s, followed by

Figure 34.   United States mean annual temperatures for the period 1901–1987. Data preparation is described by Quinlan et al. (1987); actual data are available in Boden et al. (1990).

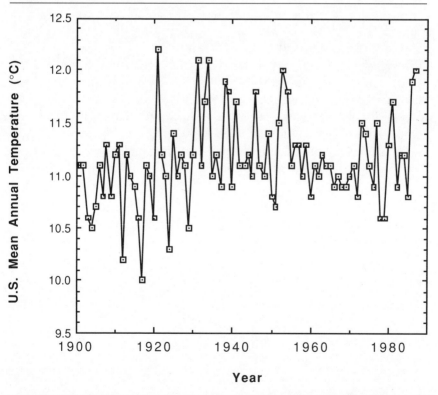

Year

a return to relatively warm years in the 1980s. Overall, the linear trend in annual temperatures is 0.29°C (0.52°F) over the 1901 to 1987 time period; however, the trend is not statistically significantly different from zero. In other words, we cannot say with confidence that there has been *any* trend in U.S. mean annual temperatures in this century! And although not statistically significant, the period from 1920 to 1987 has been dominated by a *cooling* of 0.13°C (0.24°F). Here we have possibly the best temperature data set for any area of the planet, and during a time (1920–1987) when equivalent CO_2 increased by over 30 percent (from approximately 325 to 425 ppm), the temperatures cooled slightly. If we account for remaining heat island signals and the effects of stratospheric dust, we may assume that any warming signal of the past century in the United States would be reduced even further.

The models strongly suggest that continental areas in the mid-latitudes of the Northern Hemisphere should experience most of their warming in the winter season (Mitchell et al., 1990). To test this hypothesis, the trends in season mean temperatures were computed for the HCN data. As seen in Table 5, during the summer (June–August) season, temperatures for the United States increased by 0.31°C (0.56°F) while during the winter (December–February) months, the temperatures increased by 0.47°C (0.85°F). However, these trends are not statistically significant at the 0.95 level of confidence. In addition, since 1920, summer temperatures have shown no change, but the winter temperatures have actually cooled by 0.73°C (1.32°F).

The trends in annual and seasonal temperature have shown distinctive spatial patterns across the United States (Balling and Idso, 1989).

Table 5.   Linear Changes in United States Temperature, Precipitation, and Palmer Drought Severity Index.

Season	Temperature (°C)			Precipitation (mm)	PDSI
	Min	Max	Mean		
Winter	+0.47	+0.54	+0.47	− 2.24	+0.46
Spring	+0.28	+0.46	+0.37	+11.14	+0.36
Summer	+0.38	+0.21	+0.31	− 8.40	+0.23
Autumn	+0.31	−0.19	+0.17	+29.24	+0.55
Annual	+0.35	+0.26	+0.29	+29.73	+0.40

Note: Precipitation and temperature changes are for the period 1901–1987; PDSI changes are for the period 1895–1989.

Generally, the New England area and western United States (west of the Rocky Mountains) have been warming while the central and southeastern United States have been dominated by cooling. This pattern appears in all seasons and is most pronounced in the winter season.

The HCN record also allows patterns in maximum and minimum temperatures to be examined over the United States. A plot of the mean annual minimum temperatures over this century (Figure 35) shows virtually no trend from 1901 to about 1930, a jump upward near 1930, general cooling from the early 1930s to near 1970, and warming since that time. Nonetheless, the linear trend in the minimum temperatures is statistically significant and shows an increase of 0.35°C (0.63°F).

Figure 35.   United States mean annual minimum temperatures for the period 1901–1987. Data preparation is described by Quinlan et al. (1987); actual data are available in Boden et al. (1990).

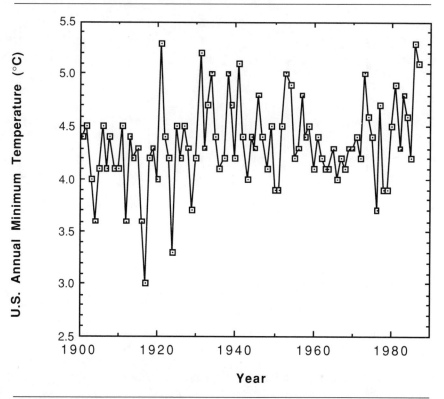

The largest portion of this warming occurs in the winter season and the least occurs in the spring (Table 5); greatest warming occurs in New England and the West, and the least amount of warming (and some cooling) occurs in the central and southeastern areas of the country. A plot of the mean annual maximum temperatures (Figure 36) shows a trend (not significant) for the 87-year period of 0.26°C (0.47°F) with strongest warming in winter and spring and cooling in autumn. Basically, the bulk of the warming seen in the United States over the past century has occurred at night and during the winter season. Small amounts of warming during the coldest time of day during the coldest months hardly make for a great disaster in the United States!

Figure 36.   United States mean annual maximum temperatures for the period 1901–1987. Data preparation is described by Quinlan et al. (1987); actual data are available in Boden et al. (1990).

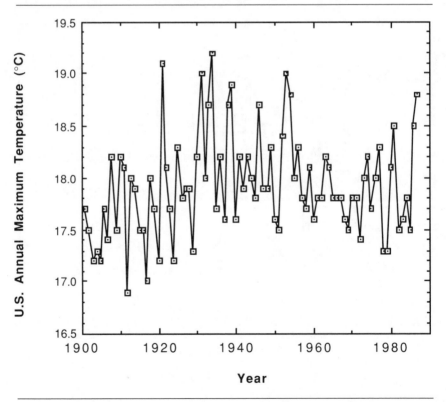

With the minimum temperatures increasing faster than the maximum temperatures, one would expect to see a reduction in the diurnal temperature range (the difference between the observed maximum and minimum temperatures). Figure 37 shows that from 1901 to about 1940, the diurnal temperature range was increasing slightly, but since that time, the range has declined steadily. This decline in United States diurnal temperature range has been noted by others (e.g., Karl et al., 1984, 1986a, 1987; Plantico et al., 1990), the same effect has been seen in other parts of the Northern Hemisphere (Folland et al., 1990), and the effect is consistent with the predictions of the models (Rind et al., 1989). The reduction in the diurnal temperature range may, in fact, be the true greenhouse signal of the past century.

Figure 37.   United States mean annual diurnal temperature range (maximum minus minimum) for the period 1901–1987. Data preparation is described by Quinlan et al. (1987).

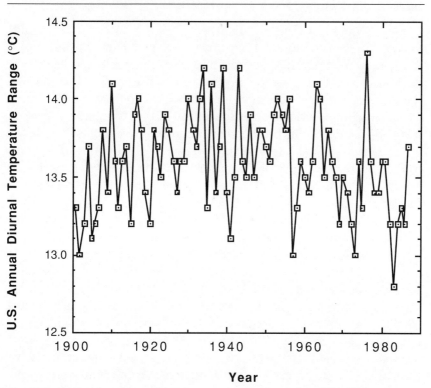

These analyses lead to several generalizations regarding temperature changes in the United States during the twentieth century. First, temperatures are increasing, but in general, the temperature increases are not statistically significant. Minimum temperatures have increased more than maximum temperatures, and the diurnal temperature range has been declining over the past 50 years. Winter is consistently the season of greatest warming while autumn tends to have the least amount of warming. Finally, warming tends to occur in the West and Northeast and cooling is the tendency in the Midwest and Southeast. The temperature record from the United States since the turn of the century is not particularly consistent with the predictions of the greenhouse models. The seasonality of the trends is broadly consistent with the models, but the magnitude of the trends is far less than what should have been observed given the buildup in the greenhouse gases.

Increases in Extreme High Temperatures?

One component of the "popular vision" is the notion that the high temperatures of the summer season will increase in magnitude and in frequency. We are told that greenhouse warming will add heat to all air masses at all times of the day, and therefore, we will see substantial increases in the frequency of record-breaking high temperatures. This prediction is not just an exaggeration of the global warming issue; it is well-founded in the professional literature. In particular, published work by Mearns et al. (1984) and Hansen et al. (1988) leads directly to the conclusion that global warming will increase the frequency of extreme high temperatures. This issue becomes especially important and threatening given the linkage between extreme high temperatures and human health, agricultural productivity, and water and energy demand.

Several empirical studies have used data from the United States to investigate the prediction of increasing extreme high temperatures. In one research effort, Sherwood Idso and I used the HCN temperature record to define areas in the United States that have had statistically significant warming or cooling over the past 40 years (Balling and Idso, 1990b). In general, we found that the summer mean maximum temperatures across the United States had *cooled* during the 1948 to 1987 study period. However, using mean summer temperatures, we identified stations in the western United States that had been warming at a significant rate and stations in New York and Texas that had been cooling at a significant rate. The daily maximum temperatures from these stations were collected for the 40-year period, and trends were established in the frequency of extremely high summer temperatures. For the stations that had experienced significant summertime warming, a 25 percent increase had occurred in the frequency of extreme events. At first glance, the prediction of increasing extreme highs with overall warming appeared to be supported. However, for the stations that had *cooled* significantly, a 38 percent *increase* in extreme high temperatures was identified. After noting that the relationships between overall warming or cooling and the frequency of extreme maximum temperatures were not statistically significant, we concluded (Balling and Idso, 1990b, p. 146) that "there is no sound observational basis for predicting an increase in the frequency of occurrence of extreme high summer temperatures in response to greenhouse warming."

In yet another research project on the maximum temperature issue, I worked with several other scientists and used a different approach to test the hypothesis that general warming will increase the frequency of extreme maximum temperatures (Balling et al., 1990). Recognizing that Phoenix temperatures were increasing at a rapid pace (Figure 14), we examined the relationship between the summer mean temperature in Phoenix and frequency of extreme maximum temperatures in the city. Our results showed that the rapid warming in Phoenix was increasing the occurrence of moderately high maximum temperatures, but somewhat surprisingly, the rapid warming was having only a small effect on the occurrence of the extreme maximum temperatures. Even the appearance of the celebrated 50°C (122°F) maximum temperature in Phoenix on June 26, 1990, did not change the statistical relation between extreme maximum temperatures and summer mean temperature. The rather simplistic scheme used by Hansen et al. (1988) for predicting massive increases in very high temperatures in United States cities would seriously overestimate the occurrence of the extreme maximum temperatures in Phoenix. We concluded (Balling et al., 1990, p. 1493) that "caution [should] be exercised in predicting distributions for extreme maximums and minimums from projected increases in mean monthly or seasonal temperatures."

PATTERNS IN U.S. PRECIPITATION

As equivalent CO_2 is doubled, the climate models tend to predict, for the conterminous United States as a whole, increases in temperature of between 2.0°C (3.6°F) and 4.0°C (7.2°F), slight increases in annual precipitation, rather substantial increases in winter precipitation, and general decreases in summer precipitation (see Mitchell et al., 1990). Given the record of the twentieth century, the empirical evidence does not strongly support the predicted increases in U.S. temperatures. Just as the HCN record allowed fairly detailed analyses of the temperature patterns, the HCN also allows careful analyses of the precipitation record.

Figure 38 presents the annual precipitation totals for the United States for the period 1901 to 1987. Between 1901 and the early 1950s, the precipitation levels varied considerably, but the average value remained stationary. In the early 1950s, a sharp drop occurred in the values, and since that time, precipitation totals have been rising at a

Figure 38. United States mean annual precipitation for the period 1901–1987. Data preparation is described by Quinlan et al. (1987).

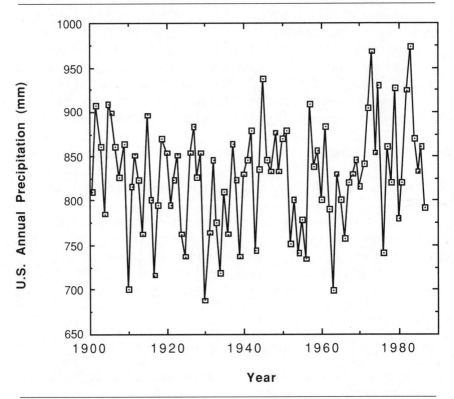

statistically significant rate. Over the entire time period (1901–1987), the precipitation levels showed a linear increase of 29.73 mm (1.17 inches) or approximately 4 percent (recall that the global precipitation 100-year increase was near 6 percent), but this increase is not statistically significant.

As seen in Table 5, the increase in precipitation from 1901 to 1987 occurs in spring and autumn while summer and winter have both shown slight decreases in precipitation. Recognizing that the numerical models predict an increase in winter precipitation (which has not occurred) and a decrease in summer rainfall (which has occurred), one would expect the ratio between winter and summer precipitation to be increasing. The plot shown in Figure 39 of the ratio of winter to summer

Figure 39.   United States precipitation ratio (winter/summer) for the period 1901–1987. Data preparation is described by Quinlan et al. (1987).

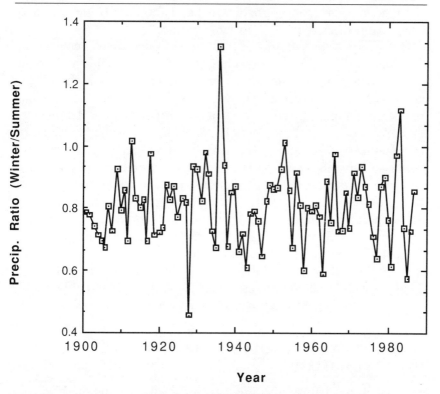

precipitation shows no trend over the entire 1901 to 1987 period, and a slight decrease since the early 1950s. Again, an expectation from the models is not seen in the empirical data.

Again working with my colleague Sherwood Idso, we found no coherent spatial patterns in precipitation trends of the United States (the same basic trends were found throughout most of the country), but we did note that contrary to model predictions, the central United States showed a much greater *increase* in precipitation than what was observed for the country as a whole (Idso and Balling, 1991b). Karl et al. (1991, p. 1059) analyzed the HCN data for the central United States and concluded that "no significant trends, even at the 0.10 significance level [that is, the 0.90 confidence level], could be found in the seasonal means of temperature, precipitation, or the ratio of winter-to-summer precipitation over the 1895 to 1989 time period." Even if the models are correct, Karl et al. (1991, p. 1058) showed that "it will likely take at least another 40 years before statistically significant precipitation changes are detected and another decade or two to detect the projected changes in temperature."

The precipitation patterns of the United States are inconclusive in resolving any part of the greenhouse debate. The observed trend appears to be toward more rainfall on an annual basis, and this observation is largely consistent with the models. Seasonal and regional trends are not particularly consistent with predictions of the models. As suggested by Karl et al. (1991), the ratio of winter to summer precipitation as a greenhouse signal should have been detected by now, and as seen in Figure 39, the ratio is trending in a direction that is inconsistent with the model predictions. To their credit, the modelers warn that the predictions of future regional precipitation patterns are highly speculative at this time. We should be very cautious in accepting the predictions of significant rainfall changes in America, particularly about the reduction of precipitation in the agricultural heartland of the country. At the present time, there appears to be little change in U.S. precipitation associated with the observed 40 percent increase in equivalent CO_2 over the past century.

PATTERNS IN U.S. DROUGHTS, CLOUD COVER, AND STREAMFLOW

Another of the many predictions associated with the "popular vision" is the increase in the frequency, magnitude, and duration of droughts in the United States, and in particular, an increase in droughts in the agricultural heartland. Like the other apocalyptic predictions, this vision of increased droughtiness is well-founded in the professional literature. Manabe et al. (1981), Manabe and Wetherald (1986, 1987), Gleick (1987), Kellogg and Zhao (1988), and McCabe et al. (1990) had all published important papers suggesting an increase in drought in the central United States as a consequence of the buildup in greenhouse gases. In possibly the most frightening paper of the series, Rind et al. (1990) showed that CO_2-induced climate effects could raise the frequency of severe droughts in the United States from 5 percent today to 50 percent by the middle of the next century.

This drought-related prediction is undoubtedly one of the most serious consequences of the greenhouse effect, and the seriousness of the prediction demands careful analysis. It is noteworthy that the predictions of increased droughtiness are far more strongly related to the increasing temperature than to any reduction in rainfall. Projected increases in temperature would raise the rate of evapotranspiration, particularly during the summer season. This increase in evapotranspiration would overwhelm any small changes in precipitation (and in addition, the precipitation totals may become more variable), and a soil moisture deficit could develop. And if a decrease in summer rainfall would occur as projected by many of the models, a significant increase in drought frequency, magnitude, and duration would certainly be realized.

Although the models generally predict this increase in drought conditions, there are many signals in the climate record that would cast doubt on the prediction. First, we have seen that the temperatures of the United States are not rising as quickly as they should to be consistent with the more apocalyptic view of the greenhouse effect. Without a large increase in temperature, the needed rise in evapotranspiration does not occur, and the droughts do not increase in magnitude, frequency, or duration. Second, we have seen a tendency for temperatures to increase at night, and not during the day. Increasing temperatures at night would have a much smaller effect on evapotranspiration than would temperature increases during the day. Third, precipitation totals are increasing on an annual basis, with largest increases in spring and autumn and only small decreases in summer rainfall. From these observed patterns in temperature and precipitation over the past century, one does not see much observational evidence to support the prediction of increased drought frequency and intensity for the United States.

An excellent drought record has been maintained for the United States throughout this century. The Palmer Drought Severity Index (PDSI) is available for some 344 climate divisions of the United States on a monthly basis. Each of these climate divisions is a relatively homogeneous area that should have similar climate conditions throughout; some of the smaller states have only one or two climate divisions while most states have five to eight divisions.

The PDSI is based on an estimate of the relative amount of soil moisture resulting from moisture inputs via precipitation and moisture outputs via evapotranspiration (Palmer, 1965). The index is standardized for each division to account for climatic differences across the country. For example, a low soil moisture level for Louisiana may be an extraordinarily high amount of soil moisture in Arizona. So the PDSI is standardized to account for "normal" patterns in some area. Values for the PDSI range from above 4.0 for extreme wetness to near 0 for near normal conditions to below −4 for extreme drought. Many climatologists (e.g., Karl and Koscielny, 1982; Diaz, 1983; Karl and Heim, 1990) have used the PDSI in their studies of historical droughts in the United States.

A plot of the annual PDSI for the United States (Figure 40) shows a highly variable pattern dominated by droughts in the early 1930s and early 1950s and relatively high moisture levels from 1970 to the mid-1980s. Over the entire 1895 to 1989 period, the trend in PDSI has been upward, but the linear increase is not statistically significant (Table 5). In fact, the PDSI values have trended upward in all seasons, but the rates of increase cannot be judged significant. Nonetheless, there is no evidence of any trend toward increasing drought as suggested by the models. Several other researchers (e.g., Karl and Heim, 1990; Idso and Balling, 1991a) have also noted the lack of any trends toward increasing drought frequency, duration, or intensity in the United States.

One may fairly argue that the models are suggesting increased drought over the central United States, and not over the United States as a whole. In a recent study of drought patterns on a state-by-state basis, Idso and I showed that the entire United States had seen a decline in PDSI values (a trend toward drought) from 1895 to the early 1950s (Idso and Balling, 1991a). However, since that time, the country had seen a reversal with a trend away from drought and toward increasing soil moisture. Of the nine states with the largest change toward

---

Figure 40.   United States Palmer Drought Severity Index (PDSI) for the period 1895–1989 (Palmer, 1965).

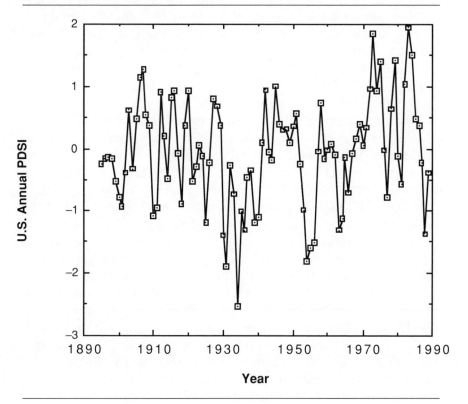

---

increasing soil moisture, six—Colorado, Iowa, Kansas, Missouri, Nebraska, and Oklahoma—are located in this agricultural heartland. Of all 48 states investigated, Nebraska showed the largest shift toward greater soil moisture availability! Once again, the predictions of the greenhouse modeling experiments are not consistent with the observations of the past century.

Several other patterns emerge from analyses of U.S. data that are strongly related to the changes in temperature and precipitation. One prediction from the greenhouse models is that due to increasing drought, lower summer rainfall, and higher evaporation rates, surface streams will show lower water levels. Karl and Riebsame (1989) examined this issue and found that fluctuations in temperature have minimal effects on streamflow, but fluctuations in precipitation are amplified by a factor of two or more. The implications for the greenhouse debate are rather obvious. The changes in temperature are likely to have no effect on streams, while the general increase in annual precipitation should increase, not decrease, streamflow. If the observed patterns of the past century are any indication of future trends, we should not accept the prediction of decreasing streamflow in a greenhouse world.

The observed increases in precipitation and the observed decrease in the diurnal temperature range would immediately suggest an increase in cloud cover over the United States. In discussing planetary changes to the climate in Chapter 4, we found that many investigators were finding increasing cloud cover over most of the world's land and ocean areas. Seaver and Lee (1987) provided one of the first comprehensive studies of U.S. cloud cover and found increasing cloudiness in the twentieth century. As with many other areas she studied, Henderson-Sellers (1989) also identified a trend to increasing cloudiness in North America. Karl and Steurer (1990) more carefully examined observing practices in the past and found cloud-cover changes prior to the 1930s and 1940s to be suspect; however, their analyses showed the upward trend in cloudiness since the 1940s to be very reliable. Plantico et al. (1990) also found that the increase in cloudiness, and particularly the increase in the autumn season, was a highly significant component of climate changes in the United States over this century. The United States, like much of the rest of the world, has become slightly warmer, wetter, and cloudier over the twentieth century.

GREENHOUSE HURRICANES IN THE UNITED STATES

The appearance of Hurricane Gilbert in the summer of 1988 fueled the notion that greenhouse warming would be associated with an increase in the frequency and intensity of hurricanes striking the United States. If Gilbert was not sufficiently convincing, Hurricane Hugo's devastation in 1989 seemed to be further proof of a connection between the greenhouse effect and hurricane activity. As with many other components of the "popular vision," the greenhouse-hurricane connection is not without solid scientific basis. For example, Emanuel (1987a) reported that a doubling of the CO_2 levels would increase sea-surface temperatures and ultimately produce a 40 to 50 percent increase in the maximum strength of hurricanes. If the sea-surface temperatures increased by 6.0°C (10.8°F), Emanuel (1988) predicted the development of extremely powerful hurricanes called hypercanes. While there may be misconceptions about Emanuel's predictions (see Idso et al., 1991), even the American Meteorological Society's policy statement issued in 1988 suggested that the greenhouse warming would probably lead to "a higher frequency and greater intensity of hurricanes" in the next 50 years (AMS Council and UCAR Board of Trustees, 1988, p. 1436).

However, not all scientists agree with the prediction of increasing hurricane activity in a greenhouse world. Idso (1989) and Idso et al. (1990) suggested that greenhouse warming may actually decrease the frequency and intensity of hurricanes. Their argument is based on two fundamental predictions from the greenhouse models. First, the models tend to predict only a small change in the temperature of the oceans in the tropical area, but the upper portion of the tropical troposphere is expected to warm much more than the underlying ocean surface. The result is less instability in the tropical atmosphere, and the greater stability would tend to retard convection and reduce the intensity of tropical cyclones. Also, the models predict greatest warming in the high latitudes and least warming in the tropics. The reduced equator-to-pole gradient in temperature would also tend to reduce the intensity and frequencies of the hurricane systems.

Recognizing the difference in predicted hurricane response to any CO_2-induced warming, Idso et al. (1990) devised an empirical experiment based on hurricane activity over the past 40 years. Within the western Atlantic, the Gulf of Mexico, and the Caribbean Sea, Idso et al. (1990) defined the number of hurricanes, the number of hurricane days, and the number of hurricanes in each of the classes defined by the widely used Saffir/Simpson hurricane intensity scale (Saffir, 1977) for each year from 1948 to 1987. The time series of the various hurricane variables were checked for trend and compared to the annual Northern Hemisphere temperature from the Jones et al. (1986a) data base.

Figures 41 and 42 show the number of hurricanes per year and the number of hurricane days per year plotted against the Northern Hemisphere temperature anomalies. In both cases, the hurricane activity variable is *inversely* related to the hemispheric temperature; although the relationship cannot be judged to be statistically significant, there is some evidence that these hurricane variables tend to decrease, and

Figure 41.   Number of hurricanes per year in the western Atlantic, Caribbean, and Gulf of Mexico vs. the Northern Hemisphere temperature anomalies for the period 1948–1987. Temperature data are from Jones et al. (1986a) and hurricane data are described by Idso et al. (1990).

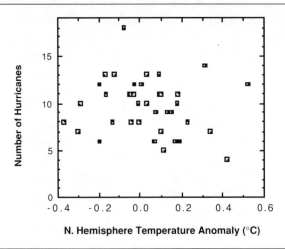

Figure 42.   Number of hurricane days per year in the western Atlantic, Caribbean, and Gulf of Mexico vs. the Northern Hemisphere temperature anomalies for the period 1948–1987. Temperature data are from Jones et al. (1986a) and hurricane data are described by Idso et al. (1990).

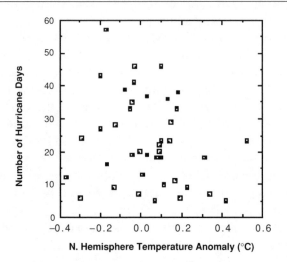

not increase, in the warmer years. Obviously, these analyses could be repeated using Atlantic and Caribbean sea-surface temperatures, but as a first approximation, the Northern Hemisphere temperature provides an adequate measure of temperature variability over a substantial portion of the globe. Idso et al. (1990, p. 262) examined a number of other hurricane activity variables and concluded that for greenhouse warming on the order of 0.5°C (0.9°F) to 1.0°C (1.8°F), "there would be no change in the frequency of occurrence of Atlantic/Caribbean hurricanes, but that there would be a significant decrease in the intensities of such storms." Their results do not support the greenhouse prediction of increased hurricane intensity and frequency in the vicinity of the United States.

THE GREENHOUSE EFFECT AND FIRES IN YELLOWSTONE PARK

Along with the many other memorable calamities of the summer of 1988, the massive wildfires in Yellowstone National Park have remained as an important part of the "popular vision." The models tend to predict much higher temperatures for the Yellowstone area and lower summer precipitation levels. The result is an increased moisture stress in the region, weakened vegetation, and a higher susceptibility to wildfires. An excellent article by Overpeck et al. (1990) shows the clear linkage between the predictions of the models and wildfire activity in the American West.

I recently worked with several other scientists on a project to determine if the model predictions and the historical records for the Yellowstone Park area are consistent (Balling et al., 1991). We began with a time series of burn area in Yellowstone National Park for each year from 1895 to 1989. The burn area values ranged from zero in 37 different years to 395,570 hectares (977,058 acres) in 1988; the year with the second largest burn area was 1910, when fires consumed 8,957 hectares (22,123 acres). The event in 1988 was clearly the event of the century! We also gathered the monthly temperature and precipitation data for the area over the same time period. Various statistical techniques were used to link the variation in the climate data to the variation in the burn area data. The result of these analyses was a composite index that could be used to quantify how favorable climate conditions were for wildfire activity.

A plot of this index for the 1895 to 1989 period shows two distinct features (Figure 43). First, and as expected, the years with the largest fires have relatively high scores on the multivariate index, suggesting a higher climate-related potential for wildfires in the years when fires were actually observed. Second, there is a statistically significant trend upward in these values, indicating a drift toward a higher fire potential. To link the observed trends with the model predictions for a doubling of CO_2, we assembled Yellowstone area temperature and precipitation predictions for $2 \times CO_2$. The model predictions of temperature and precipitation were used to generate a value of 3.87 for the multivariate fire-hazard index for a doubling of CO_2. The

Figure 43.   Plot of the Yellowstone Park climate-wildfire index for the period 1895–1989. The closed circles indicate years with the largest fires in the park (Balling et al., 1991).

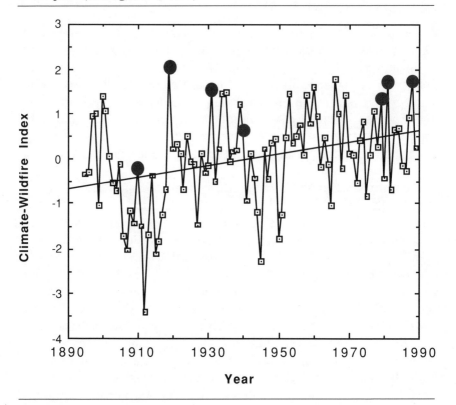

value is much higher than any value observed in the past century, but the value is reasonably consistent with the trend over the 1895 to 1989 period.

At first glance, there appears to be a solid agreement between the models and the observed trend in the climate system. However, we point out several problems in accepting the apparent connection (Balling et al., 1991). The models predict increasing temperatures in the area of approximately 4.5°C (8.1°F) and increasing precipitation totals of approximately 13 percent. The predicted aridity for a doubling of CO_2 comes from the increasing evapotranspiration rates associated with the sharp rise in temperature. The rise in evapotranspiration overwhelms the moderate increase in precipitation, and the park sees a reduction in soil moisture levels. However, the observed increase in aridity is caused by a remarkable drop in precipitation during the January through June period. Over the 1895 to 1989 period, temperatures have risen in the Park by about 0.8°C (1.4°F), but the January through June precipitation levels have decreased by 61.02 mm (2.40 inches). The models are able to predict the trend toward aridity, but the models fail to account for the trend in precipitation that is driving the trend to increasing aridity.

Obviously, wildfire activity in Yellowstone Park is only partially related to climate variations (e.g., Christensen et al., 1989), but, nonetheless, the climate trends in Yellowstone Park are undoubtedly leading to the increased aridity predicted by the models. As we have seen before in other chapters, the model predictions are not to be taken too literally at such localized scales, but in the case of Yellowstone Park, the prediction of increased aridity receives limited support in the historical climate record.

CHAPTER SUMMARY

The conterminous United States covers only 1.5 percent of the earth's surface, and therefore, what happens to the climate of the United States may seem unimportant to the debate about global climate change. And yet, the United States is a relatively large continental land mass, models make predictions for the grid points falling across the United States, an excellent set of climate records is available for the area, and some lessons can be learned when model predictions are compared to the climate patterns of the past century. Three rather important summary points stem from the analyses of the United States climate patterns:

1. Temperatures in the United States have increased over the past century, but the increase is not statistically significant from zero. We cannot say with any statistical confidence that there has been warming in the United States in the twentieth century. Most of the warming occurred prior to 1920, and in fact, from 1920 to 1987, the linear trend depicts cooling. Greatest warming has occurred in winter, and in general, more warming has occurred in the minimum temperature than the maximum temperature. The diurnal temperature range is decreasing, and there is little evidence to support any claims of increasing frequencies in extreme maximum temperatures. The seasonal and diurnal temperature patterns are broadly consistent with the model predictions given the observed increase in equivalent CO_2, but the magnitude of the warming is far below what the models suggest should have been observed.

2. The trends in precipitation for the United States are marginally consistent with the model predictions. Much like the globe as a whole, the precipitation totals have increased over the twentieth century by about 4 percent. All of the increase occurred in spring and autumn while summer and winter have shown a linear decrease in precipitation totals. The models predict an increase in the winter-to-summer precipitation ratio, but no trend in the ratio could be found in this century.

3. Virtually all of the models are predicting an increase in drought intensity, duration, and frequency in the central United States as a consequence of the buildup in the greenhouse gases. Yet, in all seasons, the trend is away from aridity and toward increasing soil moisture; the trend away from aridity is most pronounced in the central United States. Along with the observed increases in precipitation and soil moisture, cloudiness in the United States is increasing through this century.

Over the past century, and during a time when equivalent CO_2 levels rose by almost 40 percent, several signals appear which seem to be related to the exponential increase in the greenhouse gases. The conterminous United States is receiving more precipitation, cloud cover is increasing, soil moisture levels are rising, and the diurnal temperature range is decreasing. These changes have been observed throughout much of the hemisphere and globe, and these changes are largely consistent with the predictions of the models. These climate signals are not apocalyptic, but they represent what is more likely to become recognized as the true greenhouse effect.

7

AEROSOL SULFATE— THE MISSING LINK?

In our analyses of climate patterns for the globe, the hemispheres, and the United States, we found that the model predictions for the various climate elements (with the exception of temperature) are generally consistent with the observed climate trends of the past century. We saw that the accuracy of the models deteriorated as we reduced the spatial scale from the globe as a whole to a continental-scale surface. But one feature was found at all scales: the observed temperature increase of the past century was always far less than what would have been expected given the substantial increase in greenhouse gases over the past 100 years.

Climatologists have been exploring vigorously the reasons for the mismatch. The ocean thermal lag was the popular excuse for the last few years, but more recently the role of aerosol sulfate has received increased attention. In very simple terms, CO_2 and the other greenhouse gases act to warm the earth while the sulfates act to cool it (see Kaufman et al., 1991). Model experiments that use only increasing greenhouse gases as perturbations may be expected to predict increasing temperatures; if they also add the sulfates, the resulting temperature increase could be moderated. In this chapter, we will explore the possibility of bridging the greenhouse debate by exploring the role of aerosol sulfate in the atmosphere—we may find the missing link!

INTRODUCTION TO AEROSOL SULFATE

A variety of gaseous sulfur compounds are emitted into the atmosphere by a number of natural and anthropogenic sources. Major sources of these gases include sulfur dioxide (SO_2) emissions from, in decreasing order of importance, fossil fuel combustion, burning of biomass, and volcanic activity (Figure 44). Biological activity in the ocean and soil produces large quantities of dimethylsulfide that contribute significantly to the sulfur budget of the atmosphere. While there are many other aerosols in the atmosphere, the gaseous sulfur compounds represent a large portion of the total mass of the relatively small aerosols, and these sulfur compounds are particularly important in influencing the radiation balance of the earth.

The various sulfur gases undergo a photochemical conversion and form sulfuric acid (H_2SO_4) in the atmosphere. This sulfuric acid tends to condense on aerosol particles in the air, including water droplets

Figure 44.   Estimates of global emissions of gaseous sulfur compounds (from Watson et al., 1990).

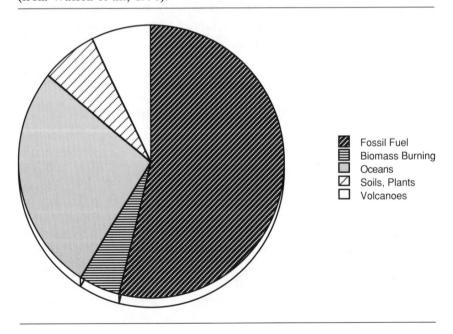

Fossil Fuel
Biomass Burning
Oceans
Soils, Plants
Volcanoes

in clouds. Evaporation leaves behind the sulfate in an aerosol phase that floats freely in the atmosphere; most of this sulfate material is located in the lower part of the troposphere. In this chapter we will be dealing with the sulfates in the lower portion of the atmosphere— earlier, we saw that sulfur compounds injected into the stratosphere by large volcanic eruptions certainly act to cool the earth by scattering incoming solar radiation.

Unlike the many greenhouse gases described in Chapter 2, the aerosol sulfates do not stay in the atmosphere for years, decades, or even centuries, but rather have an atmospheric lifetime of a few days or weeks. The particles stay relatively near the emission sources of the sulfur gases, they are carried along by the local and regional wind systems, and they tend to be rinsed quickly from the atmosphere via precipitation processes and dry fallout. This is a very important distinction between the sulfate particles and the greenhouse gases. The aerosol sulfates show tremendous variability through both time and space when compared to the greenhouse gases. Because more than 50 percent of global emission of gaseous sulfur compounds comes from burning fossil fuels, the sulfate particles have their highest concentrations in the industrialized regions of the Northern Hemisphere including eastern North America, central and eastern Europe, and eastern China. The low residence time in the atmosphere prohibits any long-range, long-term mixing of the particles throughout the planetary atmosphere (Watson et al., 1990).

In Chapter 2, we saw that the emission of CO_2 into the atmosphere shows an acceleration in the early 1950s (Figure 3); however, due to a variety of sinks for the CO_2 and due to the long residence time of CO_2, its atmospheric concentration does not show any substantial break or discontinuity in the early 1950s. The carbon budgeting of the earth-atmosphere system allows the increased CO_2 to accumulate in the atmosphere at a smooth, exponential rate through the twentieth century (Figure 4). The tropospheric mixing of the long-lived CO_2 allows concentrations to vary little from place to place around the globe.

The concentration of the sulfate particles would not be expected to show such a slow growth. Assuming the plot in Figure 3 to be representative of the volume of fossil fuel consumed, we should expect to see a rather sudden jump in anthropogenic sulfur emissions and resultant concentrations in the early 1950s. Several investigators have produced data to support this contention. Moller (1984) produced a reasonably comprehensive set of estimates of anthropogenic sulfur dioxide emissions for the globe, and his results (Figure 45) indicate an exponential increase in global SO_2 emissions over the period 1860 to 1985. Moller estimated that for the year 1985, combustion of coal would contribute 53 percent of the total emission, burning of oil would release 28 percent, lignite (brown coal) would release 11 percent, and copper smelting would account for 7 percent of the total emission.

While the overall trend in total SO_2 emission may be exponential in form, a pair of linear trends can explain more than 95 percent of the variance in the data. The first line extends from 1860 to 1950 and has a slope of 0.37 Tg S/yr (where 1 Tg or teragram is 10^{12} grams of sulfur per year). The second trend line extends from 1955 to 1985 and has a slope of 1.58 Tg S/yr; the rate of anthropogenic emission of

---

Figure 45.   Global sulfur dioxide emission (Tg S/yr). Open symbols cover the period 1860–1950; closed symbols cover the period 1955–1987. Data are from Moller (1984).

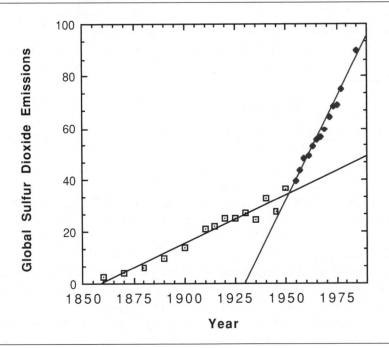

SO_2 increases by more than a factor of four in the early 1950s. Just as many of the greenhouse gases are being generated at accelerated rates, the anthropogenic emission of SO_2 is also increasing at a rapid rate.

Along with this anthropogenic increase, the release of sulfur from natural sources is also increasing. Charlson et al. (1987) postulated that biological activity in the oceanic surface waters is stimulated by increased CO_2 and higher temperatures. The increased biogenic activity results in the production of more dimethylsulfide (DMS) that is released into the atmosphere. Idso (1990b) showed that an analogous process occurs in soil microbial activity wherein increased CO_2 and temperature stimulates a rise in the metabolic rate, which in turn raises the surface-to-air flux of DMS.

Due to these increased emissions, the atmospheric concentration of sulfate has increased significantly. Mayewski et al. (1990) examined Greenland ice core data and found the mean level of non-sea-salt sulfate to be relatively stable from the early part of the century to the early 1950s. However, the sulfate loads increase significantly from the mid-1950s to the present; most recent values are more than twice as large as the values for the earlier part of the century. Watson et al. (1990) reported that estimates of sulfate concentrations in industrialized areas of eastern Europe and eastern North America are now 10 to 15 times greater than the expected natural concentration of sulfate. Hameed and Dignon (1988) showed that the greatest increase in sulfur emission between 1966 and 1980 occurred in the tropical areas, and they showed evidence for decreasing emissions in the northeastern United States and western Europe. Nonetheless, the observation of increasing sulfates in the Greenland ice core (Mayewski et al., 1990) suggests that despite the reductions in these areas, Northern Hemisphere mid-latitude sulfur emissions are increasing as a whole. The evidence is overwhelming—the concentration of aerosol sulfate has increased tremendously in most of the industrialized areas of the world, and the bulk of the increase began in the mid-1950s.

This increase in aerosol sulfate may act to cool the earth in several different ways. First, scientists have concluded that the sulfate particles act as efficient condensation nuclei and stimulate the growth of more clouds; second, the sulfate particles act to brighten existing clouds (Charlson et al., 1987; Schwartz, 1988; Albrecht, 1989; Wigley, 1989). The production of more and brighter clouds should act to reflect solar radiation back to space, reduce solar radiation reaching the surface, and ultimately cool the planet. The cooling effect should be most pronounced in the areas near, and immediately downwind of, the major industrial areas of maximum sulfur emissions. In addition to the impact on clouds, Charlson et al. (1990) showed that backscatter of sulfate particles in cloud-free air also reduces incoming solar radiation, thereby acting to cool the surface and near-surface. As we move into the 1990s, a scientific consensus is growing that the aerosol sulfates are acting to retard, to some degree, the expected warming from the buildup of greenhouse gases (e.g., Ayers et al., 1991; Foley et al., 1991; Wigley, 1991).

DETECTING THE AEROSOL SULFATE CLIMATE SIGNAL

We have seen that several spatial and temporal components in the sulfate pattern must be considered in a search for any related climate signals. First, the emission of sulfur gases by anthropogenic sources shows a sharp rise in the early 1950s. Second, the bulk of the emissions are in the industrialized Northern Hemisphere. And third, the atmospheric concentration of aerosol sulfate is greatest near the sources of SO_2 emissions. The short residence time in the atmosphere of the sulfate particles inhibits any widespread dispersion of these aerosols.

A plot of Northern Hemisphere temperature anomalies from 1891 to 1990 illustrates the potential influence of the increased aerosol sulfate loads on the hemispheric temperature (Figure 46). A trend line is established for the period 1891 to 1954 and is then extended through 1990. Notice that the trend established for the period of low sulfur emissions predicts much higher temperatures in the period of high sulfur emissions. However, only one of the 36 years in the 1955 to 1990 subperiod has a hemispheric temperature anomaly above the trend line; all of the rest have temperatures below the line. There is no guarantee that the sulfates are the cause for the departure below the line, but they certainly appear to be a legitimate candidate to explain the pattern.

A similar plot for the Southern Hemisphere (Figure 47) does not produce the same pattern. In fact, the trend established for the earlier subperiod is exceeded by 23 of the 36 years in the more recent subperiod. There appears to be nothing in the Southern Hemisphere that is causing an unusual number of negative departures away from the trend established for the 1891 to 1954 subperiod. If anything, the

Figure 46.   Northern Hemisphere temperature anomalies for 1891–1990; trend line is established for the period 1891–1954. Data are described in Jones et al. (1986a) and are available in Boden et al. (1990).

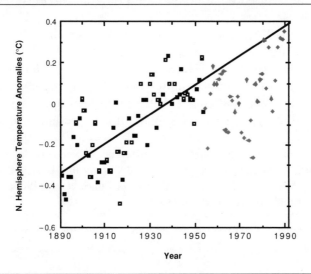

Figure 47.   Southern Hemisphere temperature anomalies for 1891–1990; trend line is established for the period 1891–1954. Data are described in Jones et al. (1986b) and are available in Boden et al. (1990).

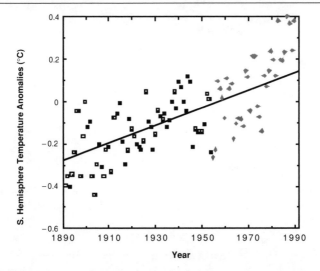

Southern Hemisphere temperatures have tended by a near two-to-one margin to exceed the predictions of the trend line. The small sulfate loading in the Southern Hemisphere appears to be unable to retard the warming caused by, among other potential forcings, the buildup in greenhouse gases.

Schwartz (1988) proposed that the differences in sulfate loadings between the hemispheres should result in a substantial cooling of the Northern Hemisphere with respect to the Southern Hemisphere during the period of maximum sulfate emissions. Idso (1990a) produced a plot of temperature differences between the hemispheres that strongly supported that hypothesis. A similar plot is presented in Figure 48

Figure 48.   Hemispheric temperature differences (Northern Hemisphere minus Southern Hemisphere) for 1891–1990; trend line is established for the period 1891–1954. Data are described in Jones et al. (1986a, 1986b) and are available in Boden et al. (1990).

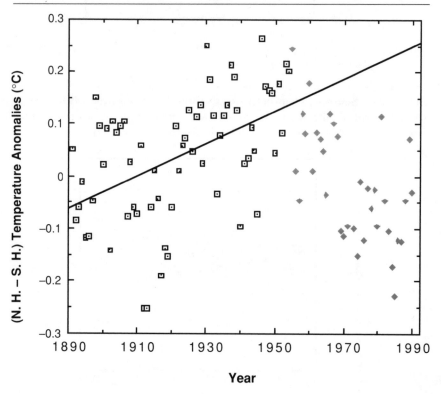

for the period 1891 to 1990. Notice that for the earlier subperiod (1891–1954), the Northern Hemisphere was warming with respect to the Southern Hemisphere. This differential warming was quite consistent with model predictions of more warming in the Northern Hemisphere compared to the Southern Hemisphere as greenhouse gases are added to the atmosphere. Rather abruptly in the 1950s, the pattern reversed, and the more recent subperiod has seen differential warming in the Southern Hemisphere (or cooling in the Northern Hemisphere). The timing of the reversal is certainly suggestive of sulfate as the cause for the observed patterns, and the magnitude of the trends suggests that the sulfate loading of the Northern Hemisphere may produce a cooling effect of approximately 1.2°C (2.2°F) per century with respect to the Southern Hemisphere (Balling and Idso, 1991c).

Recently, I worked with Sherwood Idso to develop a more stringent test for detecting a linkage between sulfate concentration and associated trends in temperature (Balling and Idso, 1991a). Our test involved plotting the sulfate loading of the atmosphere against temperature trends for various points around the globe. To illustrate this type of analysis, we selected all land-based grid points from the Jones et al. (1986a) data base for the 40°N parallel. This parallel conveniently runs through the heart of the major industrial complexes in eastern North America, eastern Europe, and eastern China. At each of these grid points with complete temperature records, the trend in mean temperatures is established for the 1955–1988 time period. Some grid points show warming while others show cooling (Figure 49). Sulfate concentrations, expressed as a ratio of total sulfur emissions (natural plus anthropogenic) to natural emissions, were then taken from a map presented by Watson et al. (1990). When the temperature trends are plotted as a function of the sulfate ratio (Figure 49), a remarkably close, negative relationship is observed. Grid points with high sulfate loadings are cooling while grid points with low sulfate loadings are warming. The analysis works remarkably well given that as a first approximation we plotted temperature trends against an index of sulfur concentrations rather than actual emissions of sulfur.

Closer inspection of Figure 49 reveals some fascinating facts about the sulfate–temperature trend relationship. The trend line suggests that a sulfate ratio of 1, indicating no anthropogenic emission of sulfur, would yield a warming of 0.14°C/decade (0.25°F/decade) at 40°N. Over the same 1955 to 1988 time period, the more pristine Southern

Figure 49.   Mean annual temperature change over the period 1955–1988 for 40°N grid points (Jones et al., 1986a) vs. the sulfate enhancement factor from Watson et al. (1990).

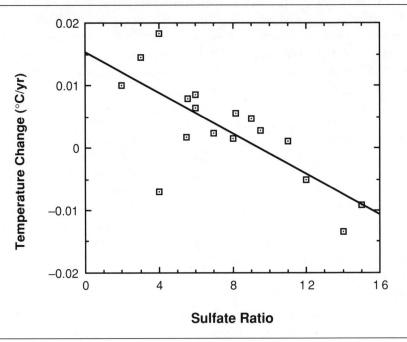

Hemisphere, with its sulfate ratios near 1.0 throughout, displays a warming of 0.12°C/decade (0.22°F/decade). Figure 49 also shows that the grid points with a sulfate ratio above 11 are cooling—the sulfate loading at these locations appears to have completely counteracted any warming forced by the buildup of greenhouse gases. These analyses provide substantial evidence that anthropogenic sulfate emissions and increases in biogenically produced sulfur gases may have a recognizable cooling effect over large parts of the globe.

Charlson et al. (1987) and many others have suggested that the increasing sulfate levels may act to produce more and brighter clouds; the cloud effects would join the backscattering of solar energy by the sulfate particles in forcing the relative cooling trends discussed above. Indeed, many studies discussed earlier (e.g., Henderson-Sellers, 1986a, 1986b, 1989; McGuffie and Henderson-Sellers, 1988; Warren et al., 1988) have suggested that cloud coverage over the land and oceans has increased in recent decades. If the planet is covered by more clouds, we could have a recognizable increase in precipitation potentially coupled to the increase in the sulfates.

A plot of the global precipitation index (Figure 50) from Diaz et al. (1989) reveals a pattern quite consistent with the trends identified for the various temperature data. The plot shows a small, statistically insignificant trend upward from 1891 to 1954. When that trend line is extended forward, we find that the global precipitation for 25 of the next 32 years is above the line with only 7 years falling below the line. The mean global precipitation index for the period 1955 to 1986 is statistically significantly higher than the mean global precipitation index for the period 1891 to 1954. Although we cannot be sure that

Figure 50.   Global precipitation index for 1891–1986 from Diaz et al. (1989). Trend line is established for the period 1891–1954.

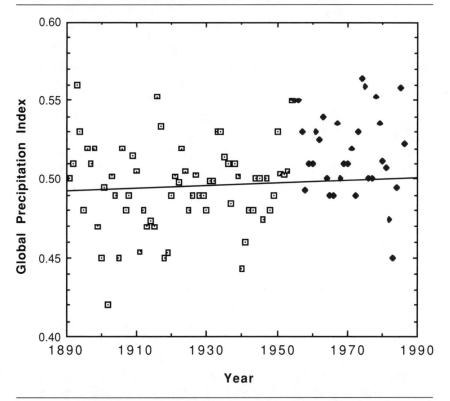

sulfates caused the increase in global precipitation, the theory and evidence suggest that sulfates have at least contributed to the increase.

SUMMARY: THE CLIMATE-SULFATE CONNECTION

In this chapter, we found that at the same time greenhouse gases were increasing exponentially in the atmosphere, sulfur dioxide and the many related aerosol sulfates were increasing as well. Some of the sulfate increase comes from increased biological activity while most comes from anthropogenic activities, including fossil fuel combustion and biomass burning. In theory, the greenhouse gases should warm the earth and the sulfates should act to cool the earth; the cooling associated with the sulfate particles comes from the brightening of clouds, the generation of new clouds, and the backscattering of solar radiation by the particles themselves. And indeed, we found that as the emission of sulfur gases increased tremendously since the mid-1950s, the following occurred:

1. Global and Northern Hemisphere temperatures did not continue to warm at the rates observed before the 1950s. Departures from the pre-established warming trends were overwhelmingly negative: the sulfates appeared to act to slow the warming trend or, in some cases, actually cause cooling to occur.
2. Using the 40°N parallel, we found that grid points with low anthropogenic sulfate loadings were warming while grid points with high anthropogenic sulfate loadings were cooling. All grid points with a sulfate ratio above 11 (a ratio of observed sulfate levels to the natural levels) displayed cooling over the recent decades.
3. The increase in the emission of sulfur gases appears to be linked with a global increase in precipitation and cloud cover. Data on cloud coverage changes are insufficient to make any firm conclusions, but the global precipitation index from land areas of the earth absolutely shows an increase after the mid-1950s.

Do sulfates resolve the greenhouse debate? Without doubt, some components of the debate are resolved, to a limited degree, by the effects of the sulfates. We found that the models were simulating a doubling of CO_2 and producing results that were largely consistent with the observations of the past century, with the obvious problem of overestimating the rise in temperature. The addition of the sulfates helps to resolve the unusual trends in the industrialized Northern Hemisphere, the lack of warming in many industrialized areas, the higher than expected precipitation totals and cloud coverage, and the increasing soil moisture levels. The sulfates do not completely resolve the debate, but future models that include both the increase in greenhouse gases and the increase in sulfate loadings are likely to produce results that can more accurately resolve the climate patterns of the past century.

8

IMPLICATIONS FOR RESPONSIBLE POLICY

So far, this book has dealt exclusively with climate, climate change, and climate response to the buildup of greenhouse gases and aerosol sulfates. The book has been written by a climatologist and not by an economist, political scientist, ecologist, or policy analyst. Nonetheless, this chapter will deal with implications of the climate reality for the construction of responsible policy. Indeed, entire books on the greenhouse effect have been written for decisionmakers (e.g., Houghton et al., 1990; Lashof and Tirpak, 1990; Synthesis Panel, 1991) and an enormous literature on policies for coping with the greenhouse world can be found in virtually any library. Decisionmakers should be able to read the first seven chapters of this book and determine how to integrate the empirical evidence into their policies. Climatologists should not be asked or forced to conduct the work of the decision-makers. The level of training required to become a responsible climatologist must be similar to the level of training required to become a responsible decisionmaker, and climatologists certainly would not ask decisionmakers to perform climatological research!

One should realize that the impact players in the greenhouse debate are constantly asked by others to comment on the implications of their work for the development of policy. What follows is a group of climate-based considerations for those who are charged with generating policy on the greenhouse issue. These are not specific recommendations for

133

specific policies, but rather, they are lessons and observations of a climatologist who from time to time is forced to think about the linkage between climate research and policy development.

CONSIDER A KINDER, GENTLER GREENHOUSE EFFECT

We started the book by examining the "popular vision" of the greenhouse effect with its apocalyptic view of the future. Rapidly rising temperatures, melting icecaps, sea-level rise, droughts in the American Great Plains, and hypercanes were all included in the rather catastrophic set of predictions for the immediate future. Yet, as we have seen, there is a large amount of empirical evidence suggesting that the apocalyptic vision is in error and that the highly touted greenhouse disaster is most improbable.

The empirical evidence suggests that our future will see a rise in temperature of approximately 1.0°C (1.8°F), with most of that warming occurring at night, in higher latitudes, and in the winter season. Extreme high temperatures probably will not increase in the $2 \times CO_2$ world, and the earth will probably be wetter, cloudier, with substantial increases in soil moisture; droughts may diminish in frequency, duration, and intensity. These predictions for the future are largely consistent with the models and largely consistent with how the climate system has already responded to a 40 percent increase in equivalent CO_2 over the past century. From a climate perspective, we have had a far more moderate greenhouse effect than what the public has been led to believe.

Decisionmakers must realize that there are legitimate reasons to believe in the apocalypse, and there is a chance that we may still be headed for the disaster. Scientists may be underestimating the thermal inertia of the oceans, the climate system may one day jump to a new, very much warmer equilibrium, or some cloud feedback or biological response may exacerbate global warming. There could be surprises, and we may fully realize the apocalypse. However, decisionmakers must be aware that there is at least an equal chance, if not a much higher chance, that many of the greenhouse effects will not be disastrous to social, biological, or economic systems.

Within the scientific community, it is a very difficult thing to label a consequence of the greenhouse effect as beneficial. Many people, including the scientists, have an immediate and knee-jerk reaction to the concept of any beneficial greenhouse effects. And yet, any change to the system will certainly create a strain on some components and some benefits to other components, and decisionmakers must become more aware of these potential benefits of the greenhouse effect. We have been trained to think only of the costs of human impacts on the environment, and yet in the case of the greenhouse effect, we can find legitimate benefits from our environmental impact.

Many examples of potential benefits of the greenhouse effect appear in the scientific literature, and many have been addressed in this book. For example, we have found some evidence that the United States is receiving more precipitation now than in the past, and that soil moisture levels are generally increasing. While there are problems with more water falling from the sky, most effects are beneficial. Also, we have found that the diurnal temperature range is decreasing in many parts of the Northern Hemisphere, a change that tends to decrease the thermal stresses at both the warm and cold extremes. And again, while there are any number of problems associated with decreasing diurnal temperature ranges, we should not overlook the potential benefits. Warmer winter temperatures in the most extreme cold Arctic air masses and more cloudiness in desert locations all carry undeniable costs and benefits, and decisionmakers must at least consider both sides of the spectrum.

In addition to the direct costs and benefits associated with the *climate* consequences of the greenhouse effect, many scientists have examined the potential benefits to the biosphere of increasing CO_2 levels. Idso (1989, p. 67) stated:

> For about a century now, and as a result of well over a thousand laboratory and field experiments, scientists have known that increasing the CO_2 content of the air around a plant's leaves nearly always leads to a significant increase in vegetative growth and development. Indeed, the plant scientific literature is replete with reviews and analyses of the subject; and fully fifty years ago, thousands of commercial nurseries were already enriching greenhouse air with extra CO_2 for the purpose of enhancing crop production, as they continue to do today. . . . [P]lants grown in air enriched with CO_2 are generally larger than similar plants grown in ambient air. They are usually taller, have more branches or tillers, more and thicker leaves containing greater amounts of chlorophyll, more and larger flowers, more and larger fruit, and more extensive root systems.

Idso's view on the net result of the buildup of greenhouse gases is highly controversial in the scientific community; he tends to see far more benefits in doubling CO_2 than costs, but his arguments are not without scientific basis.

Idso and many others (see Idso, 1989, for an extensive review) have shown that not only will increased CO_2 stimulate bigger plants, but the harvestable yield of agricultural plants should increase by an average of 33 percent as the CO_2 of the atmosphere is doubled. In addition, the increased CO_2 stimulates a partial closure of the stomatal pores of plant leaves, and as a result the plants do not transpire as much water into the atmosphere. Ultimately, the CO_2 enrichment substantially increases the water-use efficiency of plants—they produce more and they use less water. Idso (1989, p. 69) asked rhetorically, "So just how good can things get?" Well obviously, we do not know how good things can get because we do not fully understand the long-term consequences of all this goodness. Insects, diseases, competition among plants, soil nutrient depletion, plant reproduction, and a host of other potential complications could, in the eyes of Idso's critics, turn Eden into the apocalypse.

Decisionmakers must learn from the example of Idso's view of global change. Some scientists consistently see catastrophe over the horizon, and they constantly dwell on the negative consequences of future change. Other scientists see great benefits in change, but they may not fully recognize the potential environmental costs. All of these people have the luxury of being scientists; they contribute to the collective knowledge in a piece by piece fashion, often one professional journal article at a time. Decisionmakers must integrate the expanding knowledge into some type of coherent and hopefully beneficial policy. With respect to the greenhouse issue, there is a strong tendency to focus on the potential environmental costs and nearly neglect potential benefits of global climate change. Clearly, responsible policy must maximize the potential benefits and minimize the potential environmental costs associated with the ongoing buildup of the greenhouse gases.

STOPPING GLOBAL WARMING?

The "popular vision" obviously presents a rather pessimistic view of the future of the planet in the absence of some significant changes to our behavior. But according to this view of the future, if we adopt the correct set of policy options, we can "stop global warming." The theme of stopping global warming is inevitably presented as the good news in such an otherwise gloomy mess. But the evidence suggests that we cannot *stop* global warming at all, and policymakers must avoid the emotion and look closely and objectively at the climate conse-quences of their policy decisions.

The "Policymakers Summary" of the Intergovernmental Panel on Climate Change (IPCC) report (Houghton et al., 1990) provides a series of examples of how various policy scenarios could affect the rise in global temperature. In defining the various scenarios, Houghton et al. (1990, p. xxxiv) stated:

> In the *Business-as-Usual scenario* (Scenario A) the energy supply is coal intensive and on the demand side only modest efficiency increases are achieved. Carbon monoxide controls are modest, deforestation continues until the tropical forests are depleted, and agricultural emissions of methane and nitrous oxide are uncontrolled. For CFCs the Montreal Protocol is implemented albeit with only partial participation . . . In *Scenario B* the energy supply mix shifts towards lower carbon fuels, notably natural gas. Large efficiency increases are achieved. Carbon monoxide controls are stringent, deforestation is reversed, and the Montreal Protocol implemented with full participation.

By 2040, the estimate of equivalent CO_2 for Business-as-Usual is 658 ppm while the estimate of equivalent CO_2 for Scenario B is 572 ppm. Obviously, the estimated payoff for adopting and implementing the policies of Scenario B is a projected savings of 86 ppm of equivalent CO_2. However, if Scenario B is followed throughout the twenty-first century, the 658-ppm level is realized in about 2070. So in one respect, these policies are only delaying the time when we achieve various levels of equivalent CO_2.

The estimated global temperature increase associated with the Business-as-Usual and Scenario B options are present in Figure 51. If the Business-as-Usual path is followed, the estimate of planetary temperature rise is approximately 2.3°C (4.1°F) by 2040. However, if we accept Scenario B, a global warming of 1.9°C (3.4°F) is realized by 2040. The adoption of Scenario B does not *stop* global warming, but it does have the effect of *slowing* or delaying the warming. Scenario B must be presented to the public and decisionmakers as a method for *slowing* the global warming, not stopping it. *Despite all the optimism*

Figure 51.   Business-as-Usual global temperature increase (open symbols) and IPCC Scenario B temperature increase (closed symbols) based on a rise above temperatures in 1765; all data are from Houghton et al. (1990).

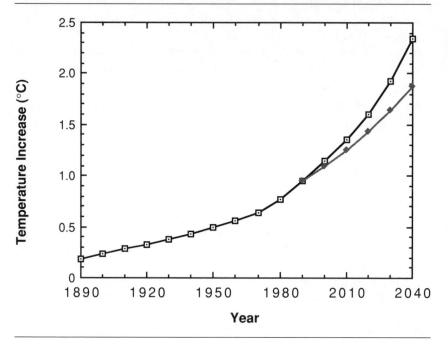

suggested by the policies of Scenario B, most of the Business-as-Usual warming is realized anyway!

A slightly different picture is revealed when the theme of this book is linked to the climate impact of policy. Figure 51 shows a warming of 0.9°C (1.6°F) for the period 1890 to 1990; yet, as we saw in Chapter 4, the actual amount of warming is exactly half that amount. Based on evidence of the past 100 years, the rate of temperature increase suggested by Figure 51 is too high by a factor of two. If we assume that this proportional error extends through to 2040, we may conclude that instead of Scenario B sparing the planet 0.4°C (0.7°F) of warming, Scenario B would only spare us half that amount. Here, the difference between Business-as-Usual and Scenario B is only 0.2°C (0.4°F).

Because the warming in Figure 51 is the warming thought to be forced by the buildup of the greenhouse gases, we would have to assume that all of the warming of the past 100 years is caused by the greenhouse gases. But as we saw throughout the book, stratospheric aerosols and contaminants to the temperature record account for some of the warming. If we accept 0.2°C (0.4°F) as the amount of unexplained warming over the past decade (the global and United States temperature trends would make this value reasonable), then the temperature curves in Figure 51 would be reduced by a factor of four, and the effect of Scenario B reduces to only 0.1°C (0.2°F) by the year 2040. It is crucial to recall that the natural variability in the climate system produces global temperature fluctuations much larger than this amount; thus, we may never be able to detect the temperature response of our policies.

We could play these hypothetical games on and on, but one message is consistent throughout. The public, and presumably many decision-makers, appear to believe that various policies can *stop* global warming. In reality, the policies seem to have remarkably little effect on the warming, despite efforts made to maximize, or even inflate, the climate impact of these policies. The public, and the decisionmakers, must be fairly and objectively told just how much of the warming will be spared by adopting various policies. People may support policies that "stop global warming," but they may be far less willing to support policies that reduce the warming by only a few tenths of a degree.

PERCEPTION VERSUS THE REALITY

The public and the decisionmakers are provided with an enormous amount of information about the possibility of a greenhouse apocalypse. Apparently, people begin to think about warming and the potentially disastrous consequences, they see many reports about record-breaking temperatures, and ultimately, their perception of warming is reinforced. People hear about the predictions of the models, they see responsible scientists who believe strongly in these predictions, and they command their policymakers to find a way to avoid the perceived catastrophe. This undoubtedly puts the policymakers in a quandary, particularly those who have been elected by a public believing in the greenhouse disaster. An elected policymaker can examine the scientific evidence and possibly conclude that the greenhouse effect may be far less harmful than is believed by constituents. With so many people believing in the greenhouse catastrophe, the elected official is confronted with telling the voters that they are wrong, even in the face of what the public may believe is mounting evidence for a catastrophe.

Unfortunately, the public perception can be remarkably different from the climate reality.

An interesting study by Harlin (1990) illustrates this point. Harlin surveyed 100 professionals (full professors, medical doctors, attorneys) in Alabama about temperature trends in their state. Fully 95 noted a significant warming trend in Alabama over the past two to four decades. Each of these 95 professionals said that the warming trend was linked to the greenhouse effect. However, Harlin analyzed the temperature trends for the state and found a distinctive warming trend from 1850 to 1935; however, a highly statistically significant cooling trend began after 1935 and continues to the present.

Here we see the problem in a nutshell. Alabama has been cooling for more than half a century, yet a survey reveals that 95 percent of the professionals in the affected area believe that a warming trend exists, one created by the greenhouse effect. These people may be concerned about the greenhouse effect raising the temperatures in their state, and they may demand policy changes to cope with continued warming. These professionals must have some influence in the state, and one could logically expect pressure on their elected officials to help build policies to address the warming issue—yet, their perception of the temperature trend is absolutely incorrect. Obviously, decisionmakers (whether elected or not) must carefully separate perceived trends from actual trends, and hopefully, the policies that result from their work will be based on hard evidence and not public perception regarding changes in the climate.

THE GREENHOUSE LEARNING CURVE

The "popular vision" is inevitably presented to the public with a sense of immediacy for action—we can stop the apocalypse if we act now. We are told that waiting for more science can push us too close to the edge of an inescapable catastrophe, and therefore, we must "do something" and we must do it immediately. However, several important points should be raised regarding the timing of our actions.

Acting in a responsible, rational fashion, we must evaluate the benefits of acting now against the penalties incurred for acting later (or not acting at all). Ausubel (1991b) and others have effectively argued that there are many reasons for waiting; future societies will be better supplied with the tools, money, knowledge, and policy options to cope with any developing greenhouse problem. Given the behavior of the climate of the past, we can certainly expect some change to occur, and the cost of cooling may far outstrip the costs of warming. The call for immediate action on the greenhouse effect "is grounded in pessimism about future technology, future economic resources, and the ability to acquire new information" (Ausubel, 1991b, p. 211).

Also, if we develop policies to cope with the perceived greenhouse threat, and we implement these policies immediately, we would initiate our actions at the current equivalent CO_2 value of approximately 430 ppm. Assume we follow the Business-as-Usual scenario for another five years and then initiate action. Based on the current trends, we would initiate our actions in 1995 in a world with about 445 ppm of equivalent CO_2. It should be noted that even if we had started some major programs in 1990, we still would be close to 445 ppm in 1995. Nonetheless, by waiting five years to "do something" we would unquestionably raise the starting point by 15 ppm of equivalent CO_2. According to the Business-as-Usual curve in Figure 51, we could expect to realize another 0.1°C (0.2°F) warming during this waiting period. For reasons discussed earlier, the plot in Figure 51 probably over-estimates the warming by at least a factor of two, and therefore, one could assume that the world may warm by about 0.05°C (0.09°F) during this period. It is noteworthy that in their objective analysis of the penalty incurred by waiting to initiate action, Schlesinger and Jiang (1991, p. 221) concluded that "the penalty is small for a 10-year delay in initiating the transition to a regime in which greenhouse-gas emissions are reduced."

Irrespective of the magnitude of the penalty in equivalent CO_2 or global warming over the next five years, one may fairly question why we should wait another year or even another minute to begin to act on the greenhouse threat. One answer comes from the examination of the Selected Bibliography of this book. Figure 52 is a plot of cumulative scientific publications on the greenhouse issue by year of the publication. As can be seen in that figure, almost half of the publications between 1800 and 1990 appeared in 1989 and 1990. If we assume that we are learning incrementally with each publication, we could conclude that our knowledge of the greenhouse effect has nearly doubled in the past few years. We are clearly in an acceleration point in the greenhouse learning curve—we are learning more about the greenhouse effect every week, and we are certain to learn a great deal more in the

Figure 52.   Cumulative frequency distribution of publications appearing in the Selected Bibliography.

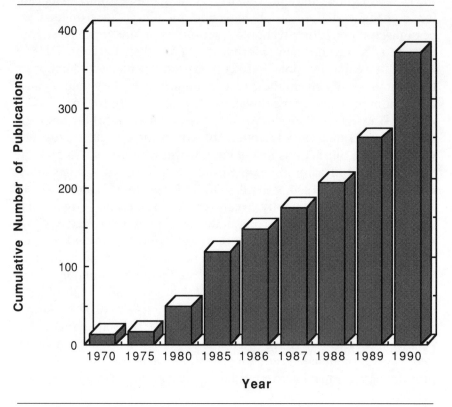

immediate future. The greenhouse effect is hot, and many scientists have been drawn to the problem, and they are producing new knowledge on the subject at an unprecedented rate. Within the next five years, we will pick up another 15 ppm in equivalent CO_2, but we will also pick up a tremendous amount of new information about the greenhouse effect.

At present, the sources and sinks of the greenhouse gases are being explored vigorously, the numerical climate models are increasing in number and in complexity, advances in computing hardware are allowing more sophisticated models to be constructed, climate data bases are being refined, new satellite data are being used to reveal climatic patterns, new impact studies are under way, and decisionmakers are probably generating improved policy options. By 1995, many of the important issues regarding the greenhouse debate will still be unresolved, but in 1995, our vision of the greenhouse effect may be substantially different than the vision of today. Policies that may make sense today may seem inappropriate five years from now. As an example, policies proposed in 1988 may not have included any consideration for the role of the aerosol sulfates; policies that made sense in 1988 may have, if implemented, actually exacerbated the magnitude of the warming (e.g., Wigley, 1991).

Like so much of the rest of the greenhouse debate, some very responsible scientists and decisionmakers believe that we must act immediately to slow the emission of greenhouse gases. Others believe that most of these actions are likely to be useless, and they see little penalty in waiting for more information on how the climate will respond to the changes in atmospheric chemistry. Irrespective of this dilemma, the scientists have an obligation to continue to explore and expose the underlying facts surrounding the greenhouse issue—*this is the time for the science and not the time for the policy implementation.* This view is very consistent with the conclusions of Schlesinger and Jiang (1991, p. 221):

> To us this small penalty does not indicate that we should "wait and see" and do nothing during this decade—quite the contrary. The study of the greenhouse effect, both theoretically and observationally, should be accelerated into a "crash programme" so that we do not squander the time that nature has given us to obtain a realistic understanding of the climate response to increasing concentrations of greenhouse gases.

The scientists are not in the business of selling fear that can be used to justify immediate policy implementation, but they are in the business of generating knowledge about a very complex climate system.

RECOGNIZE THE CLIMATE REALITY

Policy development for coping with the greenhouse issue inevitably places some stock in the outputs of the numerical climate models. In the words of NCAR climatologist Stephen Schneider, the models are the "crystal balls" that provide an imperfect glimpse into the future. Even with full knowledge of the limitations of models, decisionmakers are given some information about the future conditions; this information may then be used in various impact studies or the development of policy to avoid the negative consequences of the predicted changes. Just as policymakers are anxious to see into the future via the climate models, a closer examination of past events should also be useful in establishing policy or establishing the need to develop some policy.

A recent article by Joel Smith of the United States Environmental Protection Agency (EPA) illustrates the potential problem in not using the climate records in conducting various impact studies and presenting policy recommendations. Smith (1991) summarized a number of EPA-sponsored studies on how future climate change will affect the Great Lakes. The investigators in these studies used the $2\times CO_2$ predictions of temperature and precipitation for the Great Lakes grid points. Using three different models, they established an average rise in annual temperature of 4.6°C (8.3°F) for the region with an average rise in annual precipitation of 113 mm (4.45 inches). Based on these projected changes in climate, Smith (1991, p. 21) stated:

> These studies found that a doubling of carbon dioxide concentrations in the atmosphere could eventually lower Great Lakes water levels by 0.5 to 2.5 m; reduce ice cover by 1 to 2½ months; lengthen shipping seasons while increasing shipping and dredging costs; reduce dissolved oxygen levels in shallow lake basins; and increase fish productivity. Measures should be taken in the near future to anticipate many of these impacts and mitigate their effects or avoid costly political issues.

After reading this article, I immediately drafted a response that was later published in the same professional journal (Balling, 1991b). I noted that throughout his summary article, Smith recognized that the climate models do not produce very reliable estimates of regional changes in climate, and therefore, some care should be used in interpreting the model outputs for the specific Great Lakes area. Ironically, in the same issue of the *Bulletin of the American Meteorological Society,* the AMS (1991, p. 58) published its policy statement in which it stated that "today's climate models provide little or no useful and consistent information on regional distributions of climate change. . . . Additionally, the models are not now capable of predicting changes in average precipitation on a regional or otherwise localized basis." Obviously, the decisionmakers concerned with issues in the Great Lakes region must be made fully aware of the reliability of the predictions from the climate models; unfortunately, the reliability is very low at this localized spatial scale.

Noticeably missing in Smith's review was any mention of how the Great Lakes climate responded to the 40 percent increase in equivalent CO_2 over the past century. Figure 53 shows the temperature anomalies for the Great Lakes area from the Jones et al. (1986a) data base. Over the entire 1891 to 1988 period, temperatures have increased linearly by 0.28°C (0.50°F), but the increase is not statistically significant. However, over the past 50 years (1939 to 1988), temperatures have cooled by 0.23°C (0.41°F) while over the past 40 years (1949–1988), temperatures in the Great Lakes area have *cooled* by 0.59°C (1.06°F); these cooling trends, too, are not statistically significant. Noticeably missing are any record-breaking temperatures in the 1980s.

Figure 53.   Temperature anomalies for a Great Lakes area grid point for the period 1891–1988 (Jones et al., 1986a).

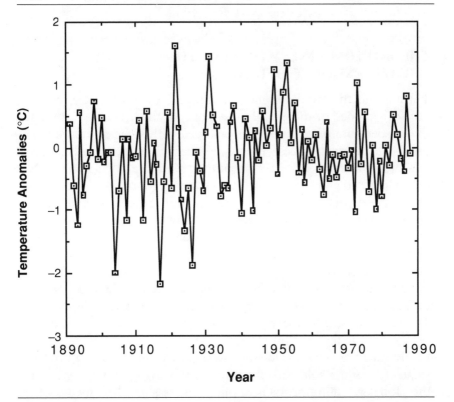

I argued that the observed trends, or lack of trends, in the Great Lakes area over the past 100 years may provide no information about the climate of the next 50 years. But while we arm the policymakers with information about projected temperature and precipitation increases from numerical modeling experiments, it is wise to also arm the decisionmakers with results from analyses of the historical climate record. Decisionmakers should be interested to learn that during a time when equivalent CO_2 increased by 40 percent, the Great Lakes area showed only a slight and insignificant increase in temperature, and that during the last half-century when equivalent CO_2 increased from 350 ppm to over 430 ppm (a 27 percent increase), the area has shown a tendency for cooling. It is also noteworthy that Assel (1990) reported that winter ice cover on the lakes has increased in extent and duration over the period 1940 to 1983. Hopefully, climatologists can provide all sorts of climate information to the policymakers, who then must decide how to proceed given both the climate model predictions and the climate patterns of the past century.

THE BOTTOM LINE: WHY LINK TO
THE GREENHOUSE DEBATE?

Throughout this book, we have examined what some scientists call "the other side" of the greenhouse story. In the first chapter, we saw that a spectrum of opinions has emerged with some reputable scientists believing in an apocalypse while other reputable scientists believe in a more moderate, less threatening greenhouse effect. This debate is not likely to end tomorrow, and as the greenhouse issue continues to receive media attention, the public and the decisionmakers will become even more aware of the conflicting views on the issue. Complicating the problem is the fact that we may need decades before a climate signal of any strength can be unequivocally linked to the greenhouse effect.

Unfortunately, any number of otherwise valuable and beneficial policies have been linked rather needlessly to the greenhouse issue. The greenhouse effect seems to present a sufficiently severe threat to demand immediate action. There has been a tendency in some cases to embrace the greenhouse effect as justification for acting now, and some policies are sold with the promise of solving or helping to solve the global warming problem. Policymakers should be aware of the developing greenhouse debate, and as the public learns more about the uncertainties surrounding the issue, they may be reluctant to accept policies that use the greenhouse effect as a centerpiece. Yet, many policies linked to the greenhouse effect (e.g., using resources more efficiently, exploring energy alternatives) make perfect sense whether the world's temperatures are warming, cooling, or remaining unchanged!

The debate on the greenhouse effect should never be used as a mandate to ravage the resources of the earth just because no climatologist can convincingly show that greenhouse gases are producing a recognizable signal in the climate system. The debate is a normal part of climatology and a normal part of science—we see debate and disagreement in all areas of the atmospheric sciences. The climate journals are filled with disagreement and debate; without this tension, there would be little science and few scientists. The more we debate and disagree, the more we work to find the answers to the critical questions surrounding the greenhouse issue. The greenhouse debate is inevitable, healthy, and generally very productive; the perceived greenhouse threat has already fostered important research on any number of interrelated social, economic, and environmental issues. Ultimately, knowledge from this debate will be used to generate responsible policy, discourage irresponsible policy, and identify the critical areas for research. In the meantime, the debate may interfere with the implementation of policy—and unquestionably, some of these policies may be useful and beneficial. But decisionmakers should recognize the overall importance of the debate, assume the debate will continue well into the next century, and then decide whether or not to employ the greenhouse issue as justification for implementing a beneficial policy. They are the "decision makers," and with respect to the greenhouse effect, they certainly have some tough decisions to make in the coming years!

9

EPILOGUE

As we have seen in the preceding eight chapters, the greenhouse gases are increasing at exponential rates largely due to human activities, and the change in atmospheric chemistry should produce some changes in the climate system. According to the "popular vision," these climate effects could be disastrous to the global ecosystem, and action must be taken now to help the planet avoid the apocalypse. But over the past century, when equivalent CO_2 levels increased by 40 percent, the climate changes appear to have been much smaller than what would be expected if indeed we were headed to that catastrophe. Scientists do not have a crystal ball, and none of the climatologists knows exactly where the climate is heading. The empirical evidence suggests minor changes while some of the climate models continue to predict more extreme changes. Decisionmakers are proposing a suite of policy options, but we cannot be sure that even the implementation of the policies will produce meaningful impacts on the climate trends. The debate over these many issues will very likely continue for decades to come.

After reading this book, you may now be much more aware of the strengths and weaknesses of the many predictions of the "popular vision." Most people have one of two different reactions after learning about the arguments presented in this book. First, many people lose their attachment to previous convictions about the greenhouse issue. They recognize that the issue is far more complex than previously thought, they recognize the validity of the arguments against the apocalypse, and they are stimulated to learn more about the issue. These people seek moderate policy steps that "do a lot of good" irrespective of the potential threat of the greenhouse effect. They want good science and good policy to go hand in hand. Their reaction is the one intended in writing the book.

The other reaction is more than predictable. Some individuals are absolutely convinced from their very limited reading that we are headed for disaster via global warming. Strangely, no amount of evidence seems to shake this crowd. They appear to have a religious attachment to the issue: they read a few reports about the potential threats associated with the greenhouse effect, and they are sold. They have a set of policies that are strongly supported by the perception of a greenhouse catastrophe, and they are not going to accept anything but the threat of disaster. These people have a sense of the arguments against accepting any lessons of the climate past, and any suggestion that some greenhouse effects will be beneficial is absolutely forbidden. They care about the environment more than they care about science.

But both of these groups are quick to link the greenhouse question to any number of other environmental threats of our time. Both groups tend to ask about acid rain, the ozone hole, deforestation, desertification, water pollution, and on and on. We all tend to think of the greenhouse effect as a problem interrelated with any number of other environmental problems. Many people argue that while some weaknesses in the greenhouse theory may exist, all of the other environmental problems must provide a sufficient threat to demand immediate action.

I hope that after reading this book you have come to appreciate the complexities of the greenhouse issue, and hopefully you are willing to at least explore the complexities of the interrelated issues. Just as there is a debate surrounding the greenhouse effect, you will find equally interesting and compelling debates with the other issues. Acid rain has its own "popular vision," but just like the greenhouse effect, the acid rain issue has its own spectrum of opinions. You will quickly find that many scientists are seriously questioning the popular anthropogenic activity–acid rain–environmental disaster linkage. Most of us are unaware of the fundamental findings of the decade-long National Acid Precipitation Assessment Program (NAPAP). After 10 years of study, the involvement of over 2,000 scientists, and an expenditure of over $500 million, the NAPAP conclusion was that an acid rain problem exists, but it is not as serious or urgent as many had feared. The answers to some tough questions on acid rain turned out to be very different from the conceptions held a decade before. Based on those answers generated from a decade of scientific inquiry, a wide range of credible options was opened. We now have a much better scientific foundation for building national and international policies to cope with the acid rain issue.

In a similar way, a tremendous amount of research is being conducted at present on the ozone hole. There is a "popular vision" for the ozone question, there is the science on the ozone hole, and there is the inevitable spectrum of opinions. Some see a disaster in the ozone layer, others see serious errors with the CFC–ozone destruction–increased ultraviolet radiation scenario. You will even find those scientists who identify benefits in reduced ozone and increased ultraviolet radiation. The same pattern can be identified for virtually any of the environmental issues that are interconnected to the greenhouse effect.

But one pattern is fairly typical in all of these issues. In the mid-1970s, acid rain emerged as the environmental threat capable of enormous destruction. After years of research, a more moderate view of the threat has been accepted by many atmospheric scientists. In the mid to late 1970s, we seemed headed for a disaster via global cooling. Clearly, the extreme predictions for catastrophe were not realized, and with sufficient research, we came to understand some of the causes of the observed cooling of the day. For a brief period in the early 1980s, climatologists became fascinated with the extreme cooling of nuclear winter. But once again, following a period of intensive research, most scientists concluded that the early predictions most likely overestimated the problem. In a similar way, some reputable scientists immediately declared that the oil fires in Kuwait would lead to climate chaos; but once again, the research on the issue moderated our view.

But all of these issues are paled by the mass and momentum of the greenhouse effect. This threat seems more real than the others, and an amazing number of scientists and decisionmakers are legitimately concerned about climate change in the coming decades. If the greenhouse effect follows the pattern of the other issues that have been addressed by climatologists over the past two decades, we can safely bet on a much more moderate change than what was originally believed and sold as the "popular vision." Only time will tell.

SELECTED BIBLIOGRAPHY

Adams, R. M., C. Rosenweig, R. M. Peart, J. T. Richie, B. A. McCarl, J. D. Glyer, R. B. Curry, J. W. Jones, K. J. Boote, and L. H. Allen, Jr., 1990: Global climate change and U.S. agriculture. *Nature,* **345,** 219–224.

Agee, E. M., 1991: Trends in cyclone and anticyclone frequency and comparison with periods of warming and cooling over the Northern Hemisphere. *Journal of Climate,* **4,** 263–267.

Albrecht, B. A., 1989: Aerosols, cloud microphysics, and fractional cloudiness. *Science,* **245,** 1227–1230.

American Meteorological Society, 1991: Policy statement of the American Meteorological Society on global climate change. *Bulletin of the American Meteorological Society,* **72,** 57–59.

American Meteorological Society Council and UCAR Board of Trustees, 1988: The changing atmosphere—challenge and opportunities. *Bulletin of the American Meteorological Society,* **69,** 1434–1440.

Andreae, M. O., and D. S. Schimel (editors), 1989: *Exchange of Trace Gases between Terrestrial Ecosystems and the Atmosphere.* New York: John Wiley & Sons.

Angell, J. K., 1986: Annual and seasonal global temperature changes in the troposphere and low stratosphere. *Monthly Weather Review,* **107,** 1922–1930.

———1988: Variations and trends in tropospheric and stratospheric global temperatures, 1958–87. *Journal of Climate,* **1,** 1296–1313.

———1990: Variation in global tropospheric temperature after adjustment for the El Niño influence 1958–89. *Geophysical Research Letters,* **17,** 1093–1096.

153

Angell, J. K., and J. Korshover, 1983: Global temperature variations in the troposphere and stratosphere. *Monthly Weather Review,* **111,** 901–921.
———1985: Surface temperature changes following the six major volcanic episodes between 1780 and 1980. *Journal of Climate and Applied Meteorology,* **24,** 937–951.
Aristarian, A. J., J. Jouzel, and C. Lorius, 1990: A 400 years isotope record of the Antarctic Peninsula climate. *Geophysical Research Letters,* **17,** 2369–2372.
Arking, A., 1991: The radiative effects of clouds and their impact on climate. *Bulletin of the American Meteorological Society,* **71,** 795–813.
Arnfield, A. J., 1987: Greenhouse effect, *in* Oliver, J. E., and R. W. Fairbridge (editors), *The Encyclopedia of Climatology.* New York: Van Nostrand Reinhold, 463–465.
Arrhenius, S., 1896: On the influence of carbonic acid in the air upon the temperature of the ground. *Philosophical Magazine,* **41,** 237–276.
Assel, R. A., 1990: An ice-cover climatology for Lake Erie and Lake Superior for the winter seasons 1897–1898 to 1982–1983. *International Journal of Climatology,* **10,** 731–748.
———1991: Implications of CO_2 global warming on Great Lakes ice cover. *Climatic Change,* **18,** 377–395.
Ausubel, J. H., 1991a: Does climate still matter? *Nature,* **350,** 649–652.
———1991b: A second look at the impacts of climatic change. *American Scientist,* **79,** 210–221.
Ayers, G. P., J. P. Ivey, and R. W. Gillett, 1991: Coherence between seasonal cycles of dimethyl sulfide, methanesulphonate and sulfate in marine air. *Nature,* **349,** 404–406.
Bacastow, R. B., C. D. Keeling, and T. P. Whorf, 1985: Seasonal amplitude increase in atmospheric CO_2 concentration at Mauna Loa, Hawaii, 1959–1982. *Journal of Geophysical Research,* **90,** 10,529–10,540.
Bach, W., 1984: CO_2-sensitivity experiments using general circulation models. *Progress in Physical Geography,* **8,** 583–609.
Baker, D. G., and D. L. Ruschy, 1989: Temperature measurements compared. *The State Climatologist,* **13,** 2–5.
Baldwin, B., J. B. Pollack, A. Summers, O. B. Toon, C. Sagan, and W. Van Camp, 1976: Stratospheric aerosols and climatic change. *Nature,* **263,** 551–555.
Balling, R. C., Jr., 1988: The climatic impact of a Sonoran vegetation discontinuity. *Climatic Change,* **13,** 99–109.
———1989: The impact of summer rainfall on the temperature gradient along the United States–Mexico border. *Journal of Applied Meteorology,* **28,** 304–308.

————1990: Testing the greenhouse gospel. *World Climate Change Report,* **1,** 30–31.

————1991a: Impact of desertification on regional and global warming. *Bulletin of the American Meteorological Society,* **72,** 232–234.

————1991b: Comments on "The potential impacts of climate change on the Great Lakes." *Bulletin of the American Meteorological Society,* **72,** 833–834.

Balling, R. C., Jr., and S. W. Brazel, 1986a: "New" weather in Phoenix? Myths and realities. *Weatherwise,* **39,** 86–90.

————1986b: Temporal analysis of summertime weather stress levels in Phoenix, Arizona. *Theoretical and Applied Climatology,* **36,** 331–342.

————1987: Time and space characteristics of the Phoenix urban heat island. *Journal of the Arizona-Nevada Academy of Science,* **21,** 75–81.

————1988: High-resolution surface temperature patterns in a complex urban terrain. *Photogrammetric Engineering and Remote Sensing,* **54,** 1289–1293.

————1989: High-resolution nighttime temperature patterns in Phoenix. *Journal of the Arizona-Nevada Academy of Science,* **23,** 49–53.

Balling, R. C., Jr., R. S. Cerveny, T. A. Miller, and S. B. Idso, 1991: Greenhouse warming may moderate British storminess. *Meteorology and Atmospheric Physics,* in press.

Balling, R. C., Jr., and S. B. Idso, 1989: Historical temperature trends in the United States and the effect of urban population growth. *Journal of Geophysical Research,* **94,** 3359–3363.

————1990a: Confusing signals in the climatic record. *Atmospheric Environment,* **24A,** 1975–1977.

————1990b: Effects of greenhouse warming on maximum summer temperatures. *Agricultural and Forest Meteorology,* **53,** 143–147.

————1990c: One hundred years of global warming? *Environmental Conservation,* **17,** 165.

————1991a: Do aerosol sulfates moderate global warming? *Theoretical and Applied Climatology,* under review.

————1991b: Decreasing diurnal temperature range: CO_2 greenhouse effect or SO_2 energy balance effect? *Atmospheric Research,* **26,** 455–459.

————1991c: Sulfate aerosols of the stratosphere and troposphere: Combined effects on surface air temperature. *Theoretical and Applied Climatology,* in press.

Balling, R. C., Jr., G. A. Meyer, and S. G. Wells, 1991: Climate change in Yellowstone National Park: Is the drought-related risk of wildfires increasing? *Climatic Change,* in press.

Balling, R. C., Jr., J. A. Skindlov, and D. H. Phillips, 1990: The impact of increasing summer mean temperatures on extreme maximum and minimum temperatures in Phoenix, Arizona. *Journal of Climate,* **3,** 1491–1494.

Barnett, T. P., 1983: Recent changes in sea level and their possible causes. *Climatic Change,* **5,** 15–38.

Barnola, J.-M., P. Pimienta, D. Raynaud, and Y. S. Korotkevich, 1991: CO_2-climate relationships as deduced from the Vostok ice core: A re-examination based on new measurements and on a re-evaluation of the air dating. *Tellus,* **43B,** 83–90.

Barnola, J.-M., D. Raynaud, Y. S. Korotkevich, and C. Lorius, 1987: Vostok ice core provides 160,000-year record of atmospheric CO_2. *Nature,* **329,** 408–414.

Bernal, P. A., 1991: Consequences of global change for oceans: A review. *Climatic Change,* **18,** 339–359.

Blake, D. R., and F. S. Rowland, 1988: Continuing worldwide increase in tropospheric methane. *Science,* **239,** 1129–1131.

Boden, T. A., P. Kanciruk, and M. P. Farrell, 1990: *Trends '90: A Compendium of Data on Global Change.* Oak Ridge, Tennessee: Carbon Dioxide Information Analysis Center, Environmental Sciences Division, Oak Ridge National Laboratory.

Bottomley, M., C. K. Folland, J. Hsiung, R. E. Newell, and D. E. Parker, 1990: *Global Ocean Surface Temperature Atlas.* Bracknell, England: Meteorological Office and Massachusetts Institute of Technology.

Bradley, R. S., 1988: The explosive eruption signal in northern hemisphere continental temperature records. *Climatic Change,* **12,** 221–243.

Bradley, R. S., H. F. Diaz, J. K. Eisheid, P. D. Jones, P. M. Kelly, and C. M. Goodess, 1987a: Precipitation fluctuations over the Northern Hemisphere land areas since the mid-19th century. *Science,* **237,** 171–175.

Bradley, R. S., H. F. Diaz, G. N. Kiladis, and J. K. Eischeid, 1987b: ENSO signal in continental temperature and precipitation records. *Nature,* **237,** 497–501.

Brazel, A. J., 1987: Urban climatology, *in* Oliver, J. E., and R. W. Fairbridge (editors), *The Encyclopedia of Climatology.* New York: Van Nostrand Reinhold, 889–901.

Bretherton, F. P., K. Bryan, and J. D. Woods, 1990: Time-dependent greenhouse-gas-induced climate change, *in* Houghton, J. T., G. J. Jenkins, and J. J. Ephraums (editors), *Climate Change: The IPCC Scientific Assessment.* Cambridge, England: Cambridge University Press, 173–193.

Broccoli, A. J., and S. Manabe, 1990: Can existing climate models be used to study anthropogenic changes in tropical cyclone intensity? *Geophysical Research Letters,* **17,** 1917–1920.

Bryan, K., F. G. Komro, S. Manabe, and M. J. Spelman, 1982: Transient climate response to increasing atmospheric carbon dioxide. *Science,* **215,** 56–58.

Bryan, K., and M. J. Spelman, 1985: The ocean's response to a CO_2-induced warming. *Journal of Geophysical Research,* 90, 11,679–11,688.

Bryant, E., 1987: CO_2-warming, rising sea-level, and retreating coasts: Review and critique. *Australian Geographer,* 18, 101–113.

Bryant, N. A., L. F. Johnson, A. J. Brazel, R. C. Balling, Jr., C. F. Hutchinson, and L. R. Beck, 1990: Measuring the effect of overgrazing in the Sonoran Desert. *Climatic Change,* 17, 243–264.

Bryson, R. A., 1989: Late Quaternary volcanic modulation of Milankovitch climate forcing. *Theoretical and Applied Climatology,* 39, 115–125.

Bryson, R. A., and T. J. Murray, 1977: *Climates of Hunger.* Madison, Wisconsin: University of Wisconsin Press.

Budyko, M. I., 1969: The effect of solar radiation variations on the climate of the earth. *Tellus,* 21, 611–619.

Caldeira, K., and M. R. Rampino, 1991: The mid-Cretaceous super plume, carbon dioxide, and global warming. *Geophysical Research Letters,* 18, 987–990.

Calendar, G. S., 1938: The artificial production of CO_2 and its influence on temperature. *Quarterly Journal of the Royal Meteorological Society,* 64, 223–237.

Cayan, D. R., and A. V. Douglas, 1984: Urban influences on surface temperatures in the southwestern United States during recent decades. *Journal of Climate and Applied Meteorology,* 23, 1520–1530.

Cess, R. D., 1990: Interpretation of an 8-year record of Nimbus 7 wide-field-of-view infrared measurements. *Journal of Geophysical Research,* 95, 16,653–16,657.

Cess, R. D., and 31 others, 1990: Intercomparison and interpretation of climate feedback processes in 19 atmospheric general circulation models. *Journal of Geophysical Research,* 95, 16,601–16,615.

Cess, R. D., G. L. Potter, J. P. Blanchet, G. J. Boer, S. J. Ghan, J. T. Kiehl, H. LeTreut, Z. X. Li, X. Z. Liang, J. F. B. Mitchell, J. J. Morcrette, D. A. Randall, M. R. Riches, E. Roeckner, U. Schlese, A. Slingo, K. E. Taylor, W. M. Washington, R. T. Wetherald, and I. Yagai, 1989: Interpretation of cloud-climate feedback as produced by 14 atmospheric general circulation models. *Science,* 245, 513–516.

Cess, R. D., and 32 others, 1991: Interpretation of snow-climate feed-back as produced by 17 general circulation models. *Science,* 253, 888–892.

Chan, Y.-H., and C. S. Wong, 1990: Long-term changes in amplitudes of atmospheric CO_2 concentrations at Ocean Station P and Alert, Canada. *Tellus,* 42B, 330–341.

Charlson, R. J., J. Langner, and H. Rodhe, 1990: Sulphate aerosol and climate. *Nature,* 348, 22.

Charlson, R. J., J. E. Lovelock, M. O. Andreae, and S. G. Warren, 1987: Oceanic phytoplankton, atmospheric sulfur, cloud albedo and climate. *Nature,* **326,** 655–661.

Charney, J. G., 1975: Dynamics of deserts and drought in the Sahel. *Quarterly Journal of the Royal Meteorological Society,* **101,** 193–202.

Christensen, N. L., J. K. Agee, P. F. Brussard, J. Hughes, D. H. Knight, G. W. Minshall, J. M. Peek, S. J. Pyne, F. J. Swanson, J. W. Thomas, S. Wells, S. E. Williams, and H. A. Wright, 1989: Interpreting the Yellowstone fires of 1988. *BioScience,* **39,** 678–685.

Cofer, W. R., III, J. S. Levine, E. L. Winstead, and B. J. Stocks, 1991: New estimates of nitrous oxide emissions from biomass burning. *Nature,* **349,** 689–691.

Conway, T. J., P. Tans, L. S. Waterman, K. W. Thoning, K. A. Masarie, and R. H. Gammon, 1988: Atmospheric carbon dioxide measurements in the remote global troposphere, 1981–1984. *Tellus,* **40 (B),** 81–115.

Covey, C., K. E. Taylor, and R. E. Dickinson, 1991: Upper limit for sea ice albedo feedback contribution to global warming. *Journal of Geophysical Research,* **96,** 9169–9174.

Crowley, T. J., 1990: Are there any satisfactory geologic analogs for a future greenhouse warming? *Journal of Climate,* **3,** 1282–1292.

Crutzen, P. J., 1991: Atmospheric chemistry: Methane's sinks and sources. *Nature,* **350,** 380–381.

Cubasch, U., and R. D. Cess, 1990: Processes and modeling, *in* Houghton, J. T., G. J. Jenkins, and J. J. Ephraums (editors), *Climate Change: The IPCC Scientific Assessment.* Cambridge, England: Cambridge University Press, 69–91.

Cunnington, W. M., and J. F. B. Mitchell, 1990: On the dependence of climate sensitivity on convective parameterization. *Climate Dynamics,* **4,** 85–93.

Danard, M. B., M. I. El-Sabh, and T. S. Murty, 1990: Recent trends in precipitation in eastern Canada. *Atmosphere-Ocean,* **28,** 140–145.

Del Genio, A. D., A. A. Lacis, and R. A. Ruedy, 1991: Simulations of the effect of a warmer climate on atmospheric humidity. *Nature,* **351,** 382–385.

Diaz, H. F., 1983: Drought in the United States: Some aspects of major dry and wet periods in the contiguous United States, 1895–1981. *Journal of Climate and Applied Meteorology,* **22,** 3–16.

———1986: An analysis of twentieth century climate fluctuations in northern North America. *Journal of Climate and Applied Meteorology,* **25,** 1625–1657.

———1990: A comparison of "global" temperature estimates from satellite and instrumental data, 1979–88. *Geophysical Research Letters,* **17,** 2373–2376.

Diaz, H. F., R. S. Bradley, and J. K. Eischeid, 1989: Precipitation fluctuations over global land areas since the late 1800s. *Journal of Geophysical Research,* **94,** 1195–1210.

Diaz, H. F., and R. G. Quayle, 1978: The 1976–77 winter in the contiguous United States in comparison with past records. *Monthly Weather Review,* **106,** 1393–1421.

———1980a: An analysis of the recent extreme winters in the contiguous United States. *Monthly Weather Review,* **108,** 687–699.

———1980b: The climate of the United States since 1895: Spatial and temporal changes. *Monthly Weather Review,* **108,** 249–266.

Dickinson, R. E., 1983: Land surface processes and climate-surface albedos and energy balance. *Advances in Geophysics,* **5,** 305–353.

———1989: Uncertainties of estimates of climatic change: A review. *Climatic Change,* **15,** 5–13.

Dickinson, R. E., and R. J. Ciceroni, 1986: Future global warming from atmospheric trace gases. *Nature,* **319,** 109–115.

Doake, C. S. M., and D. G. Vaughan, 1991: Rapid disintegration of the Wordie Ice Shelf in response to atmospheric warming. *Nature,* **350,** 328–330.

Dregne, H. E., 1977: Generalized map of the status of desertification of arid lands. Prepared by FAO of the United Nations, UNESCO and WMO for the 1977 United Nations Conference on Desertification.

Eden, P., 1988: Hurricane Gilbert. *Weather,* **43,** 446–448.

Ellsaesser, H. W., 1984: The climate effect of CO_2: A different view. *Atmospheric Environment,* **18,** 431–434.

———1986: Comments on "Surface temperature changes following the six major volcanic episodes between 1780 and 1980." *Journal of Climate and Applied Meteorology,* **25,** 1184–1185.

———1990: A different view of the climatic effect of CO_2—updated. *Atmosfera,* **3,** 3–29.

Ellsaesser, H. W., M. C. MacCracken, J. J. Walton, and S. L. Grotch, 1986: Global climatic trends as revealed by the recorded data. *Reviews in Geophysics,* **24,** 745–792.

Elsner, J. B., and A. A. Tsonis, 1991: Comparisons of observed Northern Hemisphere surface air temperature records. *Geophysical Research Letters,* **18,** 1229–1232.

Emanuel, K. A., 1986: An air-sea interaction theory for tropical cyclones. Part I: Steady-state maintenance. *Journal of the Atmospheric Sciences,* **43,** 585–604.

———1987a: The dependence of hurricane intensity on climate. *Nature,* **326,** 483–485.

———1987b: On the maximum intensity of hurricanes. *Journal of the Atmospheric Sciences,* **45,** 1143–1156.

————1988: The maximum intensity of hurricanes. *Journal of the Atmospheric Sciences,* **45,** 1143–1155.

Emery, K. O., and D. G. Aubrey, 1991: *Sea Levels, Land Levels and Tide Gauges.* New York: Springer-Verlag.

Erhardt, R. D., Jr., 1990: Reconstructed annual temperatures for the Gulf states, 1799–1988. *Journal of Climate,* **3,** 678–684.

Fisher, D. A., C. H. Hales, W.-C. Wang, M. K. W. Ko, and N. D. Sze, 1990: Model calculations of the relative effects of CFCs and their replacements on global warming. *Nature,* **344,** 513–516.

Flohn, H., 1977: Climate and energy: A scenario to a 21st century problem. *Climatic Change,* **1,** 5–20.

Foley, J. A., K. E. Taylor, and S. J. Ghan, 1991: Planktonic dimethylsulfide and cloud albedo: An estimate of the feedback response. *Climatic Change,* **18,** 1–16.

Folland, C. K., T. R. Karl, and K. Y. Vinnikov, 1990: Observed climate variations and change, *in* Houghton, J. T., G. J. Jenkins, and J. J. Ephraums (editors), *Climate Change: The IPCC Scientific Assessment.* Cambridge, England: Cambridge University Press, 195–238.

Folland, C. K., D. E. Parker, and F. C. Kates, 1984: Worldwide marine temperature fluctuations, 1856–1981. *Nature,* **310,** 670–673.

Foukal, P., and J. Lean, 1990: An empirical model of total solar irradiance variation between 1874–1988. *Science,* **247,** 556–558.

Fraedrick, K., 1978: Structural and stochastic analysis of a zero-dimensional climate system. *Quarterly Journal of the Royal Meteorological Society,* **104,** 461–474.

Fraser, P. J., W. P. Elliott, and L. S. Waterman, 1983: Atmospheric CO_2 record from direct chemical measurements during the 19th century, *in* Trabalka, J. R., and D. E. Reichle (editors), *The Changing Carbon Cycle: A Global Analysis.* New York: Springer-Verlag, 66–88.

Friedli, H., H. Lotscher, H. Oeschger, U. Siegenthaler, and B. Stauffer, 1986: Ice core record of $^{13}C/^{12}C$ ratio of atmospheric CO_2 in the past two centuries. *Nature,* **324,** 237–238.

Gal-Chen, T., and S. H. Schneider, 1976: Energy balance climate modeling: Comparison of radiative and dynamic feedback mechanisms. *Tellus,* **28,** 108–120.

Galloway, J. N., 1989: Atmospheric "acidification" projections for the future. *Ambio,* **18,** 161–166.

Gary, B. L., and S. J. Keihm, 1991: Microwave sounding units and global warming. *Science,* **251,** 316–317.

Gates, W. L., P. R. Rowntree, and Q.-C. Zeng, 1990: Validation of climate models, *in* Houghton, J. T., G. J. Jenkins, and J. J. Ephraums (editors), *Climate Change: The IPCC Scientific Assessment.* Cambridge, England: Cambridge University Press, 93–130.

Genthon, C., J. M. Barnola, D. Raynaud, C. Lorius, J. Jouzel, N. I. Barkov, Y. S. Korotkevich, and V. M. Kotlyakov, 1987: Vostok ice core: Climatic response to CO_2 and orbital forcing changes over the last climatic cycle. *Nature,* **329,** 414–418.

Ghan, S. J., K. E. Taylor, and J. E. Penner, 1990: Model test of CCN-cloud albedo climate forcing. *Geophysical Research Letters,* **17,** 607–610.

Gilliland, R. L., 1982: Solar, volcanic, and CO_2 forcing of recent climate changes. *Climatic Change,* **4,** 111–131.

Gleick, P. H., 1987: Regional hydrologic consequences of increases in atmospheric CO_2 and other trace gases. *Climatic Change,* **10,** 137–161.

Gloersen, P., and W. J. Campbell, 1991: Recent variations in Arctic and Antarctic sea-ice cover. *Nature,* **352,** 33–36.

Gordon, A. H., 1991: Global warming as a manifestation of a random walk. *Journal of Climate,* **4,** 589–597.

Gornitz, V., S. Lebedeff, and J. Hansen, 1982: Global sea level trend in the past century. *Science,* **215,** 1611–1614.

Graedel, T. E., and P. J. Crutzen, 1989: The changing atmosphere. *Scientific American,* **261,** 58–68.

Graetz, R. D., 1991: The nature and significance of the feedback of changes in terrestrial vegetation on global atmospheric and climatic changes. *Climatic Change,* **18,** 147–173.

Gray, W. M., 1984: Atlantic seasonal hurricane frequency: Part 1. *Monthly Weather Review,* **112,** 1649–1668.

Gribbin, J., 1976: *Forecasts, Famines and Freezes.* New York: Walker and Company.

Gribbin, J. (editor), 1978: *Climatic Change.* Cambridge, England: Cambridge University Press.

Grove, J. M., 1988: *The Little Ice Age.* London: Methuen.

Gurney, K. R., 1991: National greenhouse accounting. *Nature,* **353,** 23.

Halacy, D. S., Jr., 1980: *Ice or Fire? Can We Survive Climatic Change.* New York: Barnes and Noble Books.

Hall, D. O., H. E. Mynick, and R. H. Williams, 1991: Cooling the greenhouse with bioenergy. *Nature,* **353,** 11–12.

Hameed, S., and J. Dignon, 1988: Changes in the geographical distributions of global emissions of NO_x and SO_x from fossil-fuel combustion between 1966 and 1980. *Atmospheric Environment,* **22,** 441–449.

Hansen, J., I. Fung, A. Lacis, D. Rind, S. Lebedeff, R. Ruedy, G. Russell, and P. Stone, 1988: Global climate changes as forecast by Goddard Institute for Space Studies three-dimensional model. *Journal of Geophysical Research,* **93,** 9341–9364.

Hansen, J., D. Johnson, A. Lacis, S. Lebedeff, P. Lee, D. Rind, and G. Russell, 1981: Climate impact of increasing atmospheric carbon dioxide. *Science,* **213,** 957–966.

Hansen, J., A. Lacis, and M. Prather, 1989: Greenhouse effect of chloro-fluorocarbons and other trace gases. *Journal of Geophysical Research,* **94,** 16,417–16,421.

Hansen, J., A. Lacis, D. Rind, L. Russell, P. Stone, I. Fung, R. Ruedy, and J. Lerner, 1984: Climate sensitivity analysis of feedback mechanisms, *in* Hansen, J., and T. Takahashi (editors), *Climate Processes and Climate Sensitivity.* Washington, D.C.: American Geophysical Union, Geophysical Monograph 29, 130–163.

Hansen, J. E., and A. A. Lacis, 1990: Sun and dust versus greenhouse gases: An assessment of their relative roles in global climate change. *Nature,* **346,** 713–719.

Hansen, J., and S. Lebedeff, 1987: Global trends of measured surface air temperature. *Journal of Geophysical Research,* **25,** 13,345–13,372.

Hansen, J. E., W. Wang, and A. A. Lacis, 1978: Mount Agung eruption provides test of global climatic perturbation. *Science,* **199,** 1065–1068.

Hanson, K., G. A. Maul, and T. R. Karl, 1989: Are atmospheric "greenhouse" effects apparent in the climate record of the contiguous United States (1895–1987)? *Geophysical Research Letters,* **16,** 49–52.

Harlin, J. M., 1990: Warming trends resulting from the greenhouse effect: Fact versus perception. *Abstracts of the Association of American Geographers 86th Annual Meeting (Toronto Meeting),* 95.

Hart, M. H., 1978: The evolution of the atmosphere of the Earth. *Icarus,* **33,** 23–39.

Hartmann, D. L., and D. Doelling, 1991: On the net radiative effectiveness of clouds. *Journal of Geophysical Research,* **96,** 869–891.

Harvey, L. D. D., 1986: Effect of ocean mixing on the transient climate response to a CO_2 increase: An analysis of recent model results. *Journal of Geophysical Research,* **91,** 2709–2718.

———1989a: Managing atmospheric CO_2. *Climatic Change,* **15,** 343–381.

———1989b: Transient climatic response to an increase of greenhouse gases. *Climatic Change,* **15,** 15–30.

Haub, C., M. M. Kent, and M. Yanagishita, 1990: *World Population Data Sheet.* Washington, D.C.: Population Reference Bureau, Inc.

Hecht, A. D., 1983: Drought in the Great Plains: History of societal response. *Journal of Climate and Applied Meteorology,* **22,** 51–56.

Heim, R. R., Jr., 1988: About that drought. *Weatherwise,* **41,** 266–271.

Henderson-Sellers, A., 1986a: Cloud cover changes in a warmer Europe. *Climatic Change,* **8,** 25–52.

———1986b: Increasing cloud in a warming world. *Climatic Change,* **9,** 267–309.

———1989: North American total cloud amount variation this century. *Global Planetary Change,* **1,** 175–194.

————1991: Policy advice on greenhouse induced climatic change: The scientist's dilemma. *Progress in Physical Geography,* **15,** 53–70.

Henderson-Sellers, B., 1987: Modelling sea surface temperature rise resulting from increasing atmospheric carbon dioxide concentrations. *Climatic Change,* **11,** 349–359.

Holland, H. D., 1984: *The Chemical Evolution of the Atmosphere and Oceans.* Princeton, New Jersey: Princeton University Press.

Houghton, D. A., R. G. Gallimore, and L. M. Keller, 1991: Stability and variability in a coupled ocean-atmosphere climate model: Results of 100-year simulations. *Journal of Climate,* **4,** 557–577.

Houghton, J. T., G. J. Jenkins, and J. J. Ephraums (editors), 1990: *Climate Change: The IPCC Scientific Assessment.* Cambridge, England: Cambridge University Press.

Howard, L., 1833: *The Climate of London.* London: Harvey and Darton.

Hoyle, F., and N. C. Wickramasinghe, 1991: Ice particles and the greenhouse. *Nature,* **350,** 467.

Huybrechts, P., A. Letreguilly, and N. Reeh, 1991: The Greenland ice sheet and greenhouse warming. *Global Planetary Change,* **3,** 399–412.

Idso, S. B., 1984a: An empirical evaluation of earth's surface air temperature response to radiative forcing, including feedback, as applied to the CO_2-climate problem. *Archives for Meteorology, Geophysics, and Bioclimatology, Series B,* **34,** 1–19.

————1984b: What if increases in atmospheric CO_2 have an inverse greenhouse effect? I. Energy balance considerations related to surface albedo. *Journal of Climatology,* **4,** 399–409.

————1988a: Carbon dioxide and climate in the Vostok ice core. *Atmospheric Environment,* **22,** 2341–2342.

————1988b: The CO_2 greenhouse effect on Mars, Earth, and Venus. *Science of the Total Environment,* **77,** 291–294.

————1988c: Greenhouse warming or Little Ice Age demise: A critical problem for climatology. *Theoretical and Applied Climatology,* **39,** 54–56.

————1989: *Carbon Dioxide and Global Change: Earth in Transition.* Tempe, Arizona: IBR Press.

————1990a: Evidence in support of Gaian climate control: Hemispheric temperature trends of the past century. *Theoretical and Applied Climatology,* **42,** 135–137.

————1990b: A role for soil microbes in moderating the CO_2 greenhouse effect? *Soil Science,* **149,** 179–180.

————1991: The aerial fertilization effect of CO_2 and its implications for global carbon cycling and maximum greenhouse warming. *Bulletin of the American Meteorological Society,* **72,** 962–965.

Idso, S. B., and R. C. Balling, Jr., 1991a: United States droughtiness trends of the past century. *Agriculture, Ecosystems and Environment,* in press.

————1991b: Recent trends in United States precipitation. *Environmental Conservation,* **18,** 71–73.

————1991c: Surface air temperature response to increasing global industrial productivity trends: A beneficial greenhouse effect? *Theoretical and Applied Climatology,* **44,** 37–41.

————1991d: Evaluating the climatic effect of doubling atmospheric CO_2 via an analysis of Earth's historical temperature record. *Science of the Total Environment,* **106,** 239–242.

————1991e: U.S. temperature/precipitation relationships: Implications for future "greenhouse" climates. *Agricultural and Forest Meteorology,* in press.

Idso, S. B., R. C. Balling, Jr., and R. S. Cerveny, 1990: Carbon dioxide and hurricanes: Implications of Northern Hemispheric warming for Atlantic/ Caribbean storms. *Meteorology and Atmospheric Physics,* **42,** 259–263.

————1991: Reply to comments of Kerry A. Emanuel on "Carbon dioxide and hurricanes: Implications of northern hemispheric warming for Atlantic/ Caribbean storms." *Meteorology and Atmospheric Physics,* in press.

Idso, S. B., B. A. Kimball, M. G. Anderson, and J. R. Mauney, 1987: Effects of atmospheric CO_2 enrichment on plant growth: The interactive role of air temperature. *Agricultural Ecosystems and Environment,* **20,** 1–10.

Idso, S. B., and J. F. B. Mitchell, 1989: The search for CO_2/trace gas greenhouse warming. *Theoretical and Applied Climatology,* **40,** 101–102.

Jackson, R. D., and S. B. Idso, 1975: Surface albedo and desertification. *Science,* **189,** 1012–1013.

Jacoby, G. C., Jr., E. R. Cook, and L. D. Ulan, 1985: Reconstructed summer degree days in central Alaska and northwestern Canada since 1524. *Quaternary Research,* **23,** 18–26.

Jager, J., and W. W. Kellogg, 1983: Anomalies in temperature and rainfall during warm Arctic seasons. *Climatic Change,* **5,** 39–60.

Jenkinson, D. S., D. E. Adams, and A. Wild, 1991: Model estimates of CO_2 emissions from soil in response to global warming. *Nature,* **351,** 304–306.

Jones, M. D. H., and A. Henderson-Sellers, 1990: History of the greenhouse effect. *Progress in Physical Geography,* **14,** 1–18.

Jones, P. D., 1988: Hemispheric surface temperature variations: Recent trends and an update to 1987. *Journal of Climate,* **1,** 654–660.

————1990: Antarctic temperatures over the present century—a study of the early expedition record. *Journal of Climate,* **3,** 1193–1203.

Jones, P. D., P. Y. Groisman, M. Coughlan, N. Plummer, W.-C. Wang, and T. R. Karl, 1990: Assessment of urbanization effects in time series of surface air temperatures over land. *Nature,* **347,** 169–172.

Jones, P. D., and P. M. Kelly, 1983: The spatial and temporal characteristics of Northern Hemisphere surface air temperature variations. *Journal of Climatology,* **3,** 243–252.

Jones, P. D., P. M. Kelly, C. M. Goodess, and T. Karl, 1989: The effect of urban warming on the Northern Hemispheric temperature average. *Journal of Climate,* **1,** 285–290.

Jones, P. D., S. C. B. Raper, R. S. Bradley, H. F. Diaz, P. M. Kelly, and T. M. L. Wigley, 1986a: Northern hemispheric surface air temperature variations: 1851–1984. *Journal of Climate and Applied Meteorology,* **25,** 161–179.

Jones, P. D., S. C. B. Raper, and T. M. L. Wigley, 1986b: Southern hemispheric surface air temperature variations: 1851–1984. *Journal of Climate and Applied Meteorology,* **25,** 1213–1230.

Jones, P. D., and T. M. L. Wigley, 1990a: Global warming trends. *Scientific American,* **263,** 84–91.

———1990b: Satellite data under scrutiny. *Nature,* **344,** 711.

Jones, P. D., T. M. L. Wigley, C. K. Folland, D. E. Parker, J. K. Angell, S. Lebedeff, and J. E. Hansen, 1988: Evidence for global warming in the past decade. *Nature,* **332,** 790.

Jones, P. D., T. M. L. Wigley, and P. M. Kelly, 1982: Variations in surface air temperatures: Part 1: Northern Hemisphere, 1881–1980. *Monthly Weather Review,* **110,** 59–70.

Jones, P. D., T. M. L. Wigley, and P. B. Wright, 1986c: Global temperature variations between 1861 and 1984. *Nature,* **322,** 430–434.

Joos, F., J. L. Sarmiento, and U. Siegenthaler, 1991: Estimates of the effect of Southern Ocean iron fertilization on atmospheric CO_2 concentrations. *Nature,* **349,** 772–775.

Kalkstein, L. S. (editor), 1991: *Global Comparisons of Selected GCM Control Runs and Observed Climate Data.* Washington, D.C.: U.S. Environmental Protection Agency.

Kalkstein, L. S., P. C. Dunne, and R. S. Vose, 1990: Detection of climatic change in the western North American Arctic using a synoptic climatological approach. *Journal of Climate,* **3,** 1153–1167.

Kane, R. P., and N. R. Teixeira, 1990: Power spectrum analysis of the time-series of annual mean surface temperatures. *Climatic Change,* **17,** 121–130.

Karl, T. R., 1985: Perspective on climate change in North America during the twentieth century. *Physical Geography,* **6,** 207–229.

———1988: Multi-year fluctuations of temperature and precipitation: The gray area of climatic change. *Climatic Change,* **12,** 179–197.

Karl, T. R., H. F. Diaz, and G. Kukla, 1988: Urbanization: Its detection and effect in the United States climatic record. *Journal of Climate,* **1,** 1099–1123.

Karl, T. R., and R. R. Heim, Jr., 1990: Are droughts becoming more frequent or severe in the United States? *Geophysical Research Letters,* **17,** 1921–1924.

Karl, T. R., R. R. Heim, Jr., and R. G. Quayle, 1991: The greenhouse effect in central North America: If not now, when? *Science,* **251,** 1058–1061.

Karl, T. R., and P. D. Jones, 1989: Urban bias in area-averaged surface air temperature trends. *Bulletin of the American Meteorological Society,* **70,** 265–270.

Karl, T. R., and A. J. Koscielny, 1982: Drought in the United States: 1895–1981. *Journal of Climate,* **2,** 313–329.

Karl, T. R., G. Kukla, and J. Gavin, 1984: Decreasing diurnal temperature range in the United States and Canada, 1941 through 1980. *Journal of Climate and Applied Meteorology,* **23,** 1489–1504.

———1986a: Relationship between decreased temperature range and precipitation trends in the United States and Canada, 1941–80. *Journal of Climate and Applied Meteorology,* **25,** 1878–1886.

———1987: Recent temperature changes during overcast and clear skies in the United States. *Journal of Climate and Applied Meteorology,* **26,** 698–711.

Karl, T. R., and R. G. Quayle, 1981: The 1980 summer heat wave and drought in historical perspective. *Monthly Weather Review,* **109,** 2055–2072.

———1988: Climatic change in fact and theory: Are we collecting the facts? *Climatic Change,* **13,** 5–17.

Karl, T. R., and W. E. Riebsame, 1989: The impact of decadal fluctuations in mean precipitation and temperature on runoff: A sensitivity study over the United States. *Climatic Change,* **15,** 423–447.

Karl, T. R., and P. M. Steurer, 1990: Increased cloudiness in the United States during the first half of the twentieth century: Fact or fiction? *Geophysical Research Letters,* **17,** 1925–1928.

Karl, T. R., J. D. Tarpley, R. G. Quayle, H. F. Diaz, D. A. Robinson, and R. S. Bradley, 1989: The recent climate record: What it can and cannot tell us. *Reviews in Geophysics,* **27,** 405–430.

Karl, T. R., and C. N. Williams, Jr., 1987: An approach to adjusting climatological time series for discontinuous inhomogeneities. *Journal of Climate and Applied Meteorology,* **27,** 1744–1763.

Karl, T. R., C. N. Williams, Jr., P. J. Young, and W. M. Wendland, 1986b: A model to estimate the time of observation bias with monthly mean maximum, minimum, and mean temperatures for the United States. *Journal of Climate and Applied Meteorology,* **25,** 145–160.

Karoly, D. J., 1987: Southern Hemisphere temperature trends: A possible greenhouse gas effect? *Geophysical Research Letters,* **14,** 1139–1141.

Kaufman, Y. J., R. S. Fraser, and R. L. Mahoney, 1991: Fossil fuel and biomass burning effect on climate—heating or cooling? *Journal of Climate,* **4,** 578–588.

Keller, M., W. A. Kaplan, and S. C. Wofsy, 1986: Emissions of N_2O, CH_4, and CO_2 from tropical soils. *Journal of Geophysical Research,* **91,** 11,791–11,802.

Kellogg, W. W., 1991: Response to skeptics of global warming. *Bulletin of the American Meteorological Society,* **72,** 499–511.

Kellogg, W. W., and R. Schware, 1981: *Climate Change and Society.* Boulder, Colorado: Westview Press.

Kellogg, W. W., and Z. C. Zhao, 1988: Sensitivity of soil moisture to doubling of carbon dioxide in climate model experiments. Part I: North America. *Journal of Climate,* **1,** 348–366.

Kelly, P. M., P. D. Jones, T. M. L. Wigley, C. B. Sear, S. G. Cherry, and R. K. Tavakol, 1982: Variations in surface air temperatures: Part 2. Arctic regions, 1881–1980. *Monthly Weather Review,* **110,** 71–83.

Kelly, P. M., and T. M. L. Wigley, 1990: The influence of solar forcing trends on global mean temperature since 1861. *Nature,* **347,** 460–462.

Khalil, M. A. K., and R. A. Rasmussen, 1983: Increase and seasonal cycles in the atmospheric concentration of nitrous oxide (N_2O). *Tellus,* **35B,** 161–169.

————1990: Atmospheric methane: Recent global trends. *Environmental Science and Technology,* **24,** 549–553.

Kukla, G., J. Gavin, and T. R. Karl, 1986: Urban warming. *Journal of Climate and Applied Meteorology,* **25,** 1265–1270.

Kuo, C. C., C. Lindberg, and D. J. Thompson, 1990: Coherence established between atmospheric carbon dioxide and global temperature. *Nature,* **343,** 709–714.

Lachenbruch, A. H., and B. V. Marshall, 1986: Changing climate: Geothermal evidence from permafrost in the Alaskan Arctic. *Science,* **234,** 689–696.

Lacis, A., J. Hansen, P. Lee, T. Mitchell, and S. Lebedeff, 1981: Greenhouse effect trace gases, 1970–1980. *Geophysical Research Letters,* **8,** 1035–1038.

Lamb, H. H., 1982: *Climate History and the Modern World.* London: Methuen.

Landsberg, H. E., 1981: *The Urban Climate.* London: Academic Press.

Landsberg, H. E., and J. M. Albert, 1971: The summer of 1816 and volcanism. *Weatherwise,* **27,** 63–66.

Langley, S. P., 1886: Observations on invisible heat-spectra and the recognition of hitherto unmeasured wave-lengths, made at the Allegheny Observatory. *Philosophical Magazine,* **31,** 394–409.

Lashof, D. A., 1989: The dynamic greenhouse: Feedback processes that may influence future concentrations of atmospheric trace gases and climatic change. *Climatic Change,* **14,** 213–242.

Lashof, D. A., and D. R. Ahuja, 1990: Relative contributions of greenhouse gas emissions to global warming. *Nature,* **344,** 529–531.

Lashof, D. A., and D. A. Tirpak (editors), 1990: *Policy Options for Stabilizing Global Climate.* New York: Hemisphere Publishing Company.

Latham, J., and M. H. Smith, 1990: Effect on global warming of wind-dependent aerosol generation at the ocean surface. *Nature,* **347,** 372–373.

Lean, J., and D. A. Warrilow, 1989: Simulation of the regional climatic impact of Amazon deforestation. *Nature,* **342,** 411–413.

Lee, D. O., 1984: Urban climates. *Progress in Physical Geography,* **8,** 1–31.

Lee, R., 1973: The "greenhouse" effect. *Journal of Applied Meteorology,* **12,** 556–557.

Levander, T., 1990: The relative contributions of the greenhouse effect from the use of different fuels. *Atmospheric Environment,* **24A,** 2707–2714.

Lindstrom, D. R., and D. R. MacAyeal, 1990: Scandinavian, Siberian, and Arctic Ocean glaciation: Effect of Holocene atmospheric CO_2 variations. *Science,* **245,** 628–631.

Lindzen, R. S., 1990: Some coolness concerning global warming. *Bulletin of the American Meteorological Society,* **71,** 288–299.

Liu, Q., and C. J. E. Schuurmans, 1990: The correlation of tropospheric and stratospheric temperature and its effect on the detection of climate change. *Geophysical Research Letters,* **17,** 1085–1088.

Lorenz, E. N., 1970: Climatic change as a mathematical problem. *Journal of Applied Meteorology,* **9,** 325–329.

Lorius, C., J. Jouzel, D. Raynaud, J. Hansen, and H. Le Treut, 1990: The ice-core record: Climate sensitivity and future greenhouse warming. *Nature,* **347,** 139–145.

Lough, J. M., T. M. L. Wigley, and J. P. Palutikof, 1983: Climate and climate impact scenarios for Europe in a warmer world. *Journal of Climate and Applied Meteorology,* **22,** 1673–1684.

MacCracken, M. C., 1987: Carbon dioxide and climate, *in* Oliver, J. E., and R. W. Fairbridge (editors), *The Encyclopedia of Climatology.* New York: Van Nostrand Reinhold Company, 185–195.

MacCracken, M. C., and F. M. Luther (editors), 1985: *Projecting the Climatic Effects of Increasing Carbon Dioxide.* Washington, D.C.: United States Department of Energy.

Madden, R. A., and V. Ramanathan, 1980: Detecting climatic change due to increasing carbon dioxide. *Science,* **209,** 763–768.

Maheras, P., and F. Kolyva-Machera, 1990: Temporal and spatial characteristics of annual precipitation over the Balkans in the twentieth century. *International Journal of Climatology,* **10,** 495–504.

Manabe, S., 1983: Carbon dioxide and climate change. *Advances in Geophysics,* **25,** 39–82

Manabe, S., and R. J. Stouffer, 1980: Sensitivity of a global climate model to an increase of CO_2 concentration in the atmosphere. *Journal of Geophysical Research,* **85,** 5529–5554.

Manabe, S., and R. T. Wetherald, 1967: Thermal equilibrium of the atmosphere with a given distribution of relative humidity. *Journal of the Atmospheric Sciences,* **24,** 241–259.

————1975: The effects of doubling the CO_2 concentration on the climate of a general circulation model. *Journal of the Atmospheric Sciences,* **32,** 3–15.

————1980: On the distribution of climatic change resulting from an increase in the CO_2 content of the atmosphere. *Journal of the Atmospheric Sciences,* **37,** 99–118.

————1986: Reduction in summer soil wetness induced by an increase in atmospheric carbon dioxide. *Science,* **232,** 626–628.

————1987: Large-scale changes of soil wetness induced by an increase in atmospheric carbon dioxide. *Journal of the Atmospheric Sciences,* **44,** 1211–1235.

Manabe, S., R. T. Wetherald, and R.J. Stouffer, 1981: Summer dryness due to an increase of atmospheric CO_2 concentration. *Climatic Change,* **3,** 347–386.

Marland, G., and R. M. Rotty, 1984: Carbon dioxide emissions from fossil fuels: A procedure for estimation and results from 1950–82. *Tellus,* **36B,** 232.

Marston, J. B., M. Oppenheimer, F. K. Fujita, and S. R. Gaffin, 1991: Carbon dioxide and temperature. *Nature,* 349, 573–574.

Mass, C. F., and D. A. Portman, 1989: Major volcanic eruptions and climate: A critical evaluation. *Journal of Climate,* **2,** 566–593.

Matthews, S. W., 1976: What's happening to our climate? *National Geographic,* **150,** 576–615.

Mayewski, P. A., W. B. Lyons, M. J. Spencer, M. S. Twickler, C. F. Buck, and S. Whitlow, 1990: An ice-core record of atmospheric response to anthropogenic sulfate and nitrate. *Nature,* **346,** 554–556.

McCabe, G. J., Jr., D. M. Wolock, L. E. Hay, and M. A. Ayers, 1990: Effects of climatic change on the Thornthwaite moisture index. *Water Resources Bulletin,* **26,** 633–643.

McGuffie, K., and A. Henderson-Sellers, 1988: Is Canadian cloudiness increasing? *Atmosphere-Ocean,* **26,** 608–633.

McKibben, W., 1989: *The End of Nature.* New York: Random House.

Mearns, L. O., R. W. Katz, and S. H. Schneider, 1984: Extreme high temperature events: Changes in their probabilities with changes in mean temperature. *Journal of Climate and Applied Meteorology,* **23,** 1601–1613.

Mearns, L. O., S. H. Schneider, S. L. Thompson, and L. R. McDaniel, 1990: Analysis of climate variability in general circulation models: Comparison with observation and changes in variability in 2×CO$_2$ experiments. *Journal of Geophysical Research,* **95,** 20,469–20,490.

Meehl, G. A., 1984: Modeling the earth's climate. *Climatic Change,* **6,** 259–286.

————1990: Development of global coupled ocean-atmosphere general circulation models. *Climate Dynamics,* **5,** 19–33.

Meehl, G. A., and W. M. Washington, 1990: CO$_2$ climate sensitivity and snow-sea-ice albedo parameterization in an atmospheric GCM coupled to a mixed-layer ocean model. *Climatic Change,* **16,** 283–306.

Michaels, P. J., 1990: The greenhouse effect and global change: Review and reappraisal. *International Journal of Environmental Studies,* **36,** 55–71.

Michaels, P. J., D. E. Sappington, and D. E. Stooksbury, 1988: Anthropogenic warming in north Alaska? *Journal of Climate,* **1,** 942–945.

Mitchell, J. F. B., 1983: The seasonal response of a general circulation model to changes in CO$_2$ and sea temperatures. *Quarterly Journal of the Royal Meteorological Society,* **109,** 113–152.

————1989: The "greenhouse" effect and climate change. *Reviews in Geophysics,* **29,** 30–60.

————1990: Greenhouse warming: Is the mid-Holocene a good analogue? *Journal of Climate,* **3,** 1177–1192.

Mitchell, J. F. B., S. Manabe, V. Meleshko, and T. Tokioka, 1990: Equilibrium climate change—and its implications for the future, *in* Houghton, J. T., G. J. Jenkins, and J. J. Ephraums (editors), *Climate Change: The IPCC Scientific Assessment,* Cambridge, England: Cambridge University Press, 131–172.

Mitchell, J. F. B., C. A. Senior, and W. J. Ingram, 1989: CO$_2$ and climate: A missing feedback? *Nature,* **341,** 132–134.

Mitchell, J. F. B., C. A. Wilson, and W. M. Cunnington, 1987: On CO$_2$ climate sensitivity and model dependence of results. *Quarterly Journal of the Royal Meteorological Society,* **113,** 293–322.

Mitchell, J. M., Jr., 1953: On the causes of instrumentally observed secular temperature trends. *Journal of Meteorology,* **10,** 244–261.

————1958: Effect of changing observation time on mean temperatures. *Bulletin of the American Meteorological Society,* **39,** 83–89.

————1991: Carbon dioxide and future climate. *Weatherwise,* **44,** 17–23.

Mo, K. C., and H. van Loon, 1985: Climate trends in the Southern Hemisphere. *Journal of Climate and Applied Meteorology,* **24,** 777–789.

Moller, D., 1984: Estimation of the global man-made sulphur emission. *Atmospheric Environment,* **18,** 19–27.

Möller, F., 1963: On the influence of changes in the CO$_2$ concentration in air on the radiation balance of the earth's surface and on the climate. *Journal of Geophysical Research,* **68,** 3877–3886.

Namias, J., 1980: Some concomitant regional anomalies associated with hemispherically averaged temperature variations. *Journal of Geophysical Research,* **85,** 1585–1590.

———1989: The greenhouse effect as a symptom of our collective angst. *Oceanus,* **32,** 65–67.

National Research Council, 1983: *Changing Climate.* Washington, D.C.: National Academy Press.

Neftel, A., E. Moor, H. Oeschger, and B. Stauffer, 1985: Evidence from polar ice cores for the increase in atmospheric CO_2 in the past two centuries. *Nature,* **315,** 45–47.

Newell, N. E., R. E. Newell, J. Hsiung, and Z. Wu, 1989: Global marine temperature variation and the solar magnetic cycle. *Geophysical Research Letters,* **16,** 311–314.

Newell, R. E., 1970: Stratospheric temperature change from the Mt. Agung volcanic eruption of 1963. *Journal of the Atmospheric Sciences,* **27,** 977–978.

Nisbet, E., 1990: Climate change and methane. *Nature,* **347,** 23.

Oglesby, R. J., and B. Saltzman, 1990: Sensitivity of the equilibrium surface temperature of a GCM to systematic changes in atmospheric carbon dioxide. *Geophysical Research Letters,* **17,** 1089–1092.

Oke, T. R., 1979: *Review of Urban Climatology, 1973–1976.* Geneva, Switzerland: WMO Technical Note 169.

Overpeck, J. T., D. Rind, and R. Goldberg, 1990: Climate-induced changes in forest disturbance and vegetation. *Nature,* **343,** 51–53.

Palmer, W. C., 1965: Meteorological drought. Washington, D.C.: U.S. Weather Bureau Research Paper 45.

Paltridge, G. W., 1980: Cloud-radiation feedback to climate. *Quarterly Journal of the Royal Meteorological Society,* **106,** 895–899.

———1991: Rainfall-albedo feedback to climate. *Quarterly Journal of the Royal Meteorological Society,* **117,** 647–650.

Palutikof, J. P., T. M. L. Wigley, and J. M. Lough, 1984: *Seasonal Climate Scenarios for Europe and North America in a High-CO_2, Warmer World.* Washington, D.C.: U.S. Department of Energy.

Parker, D. E., 1985: On the detection of temperature changes induced by increasing atmospheric carbon dioxide. *Quarterly Journal of the Royal Meteorological Society,* **111,** 587–601.

Patzelt, G., 1989: Die 1980er-Vorstossperiode der Alpengletcher. *Oesterreicher Alpenverein,* **44,** 14–15.

Payette, S., L. Filion, A. Delwaide, and C. Begin, 1989: Reconstruction of tree-line vegetation response to long-term climate change. *Nature,* **341,** 429–432.

Pearce, F., 1986: How to stop the greenhouse effect. *New Scientist,* **111,** 29–30.

Pearman, G. I., D. Etheridge, F. DeSilva, and P. J. Fraser, 1986: Evidence of changing concentrations of atmospheric CO_2, N_2O, and CH_4 from air bubbles in Antarctic ice. *Nature,* **320,** 248–250.

Peel, D. A., R. Mulvaney, and B. M. Davison, 1988: Stable-isotope/air temperature relationships in ice cores from Dolleman Island and the Palmer Land Plateau, Antarctica Peninsula. *Annals of Glaciology,* **10,** 130–136.

Pitman, A. J., A. Henderson-Sellers, and Z.-L. Yang, 1990: Sensitivity of regional climates to localized precipitation in global models. *Nature,* **346,** 734–737.

Pittock, A. E., and M. J. Salinger, 1982: Towards regional scenarios for a CO_2-warmed earth. *Climatic Change,* **4,** 23–40.

Plantico, M. S., T. R. Karl, G. Kukla, and J. Gavin, 1990: Is the recent climate change across the United States related to rising levels of anthropogenic greenhouse gases? *Journal of Geophysical Research,* **95,** 16,617–16,637.

Plass, G. N., 1956: The CO_2 theory of climatic change. *Tellus,* **8,** 140–153.

Platt, C. M. R., 1989: The role of cloud microphysics in high-cloud feedback effects on climatic change. *Nature,* **341,** 428–429.

Pollack, J. B., O. B. Toon, C. Sagan, A. Summers, B. Baldwin, and W. Van Camp, 1976: Volcanic explosions and climatic change: A theoretical assessment. *Journal of Geophysical Research,* **81,** 1071–1083.

Ponte, L., 1976: *The Cooling.* Englewood Cliffs, New Jersey: Prentice-Hall.

Post, W. M., T.-H. Peng, W. E. Emanuel, A. W. King, and V. H. Dale, 1990: The global carbon cycle. *American Scientist,* **78,** 310–326.

Prentice, K. C., and I. Y. Fung, 1990: The sensitivity of terrestrial carbon storage to climate change. *Nature,* **346,** 48–51.

Quadfasel, D., A. Sy, D. Wells, and A. Tunik, 1991: Warming in the Arctic. *Nature,* **350,** 385.

Quinlan, F. T., T. R. Karl, and C. N. Williams, Jr., 1987: United States historical climatology network (HCN) serial temperature and precipitation data. Oak Ridge, Tennessee: Carbon Dioxide Information Analysis Center, Oak Ridge National Laboratory, NDP–019.

Radke, L. F., J. A. Coakley, Jr., and M. D. King, 1989: Direct and remote sensing observations of the effects of ships on clouds. *Science,* **246,** 1146–1149.

Ramanathan, V., 1981: The role of ocean-atmospheric interactions in the CO_2 climate problem. *Journal of the Atmospheric Sciences,* **38,** 918–930.

———1988: The greenhouse theory of climate change: A test by an inadvertent global experiment. *Science,* **240,** 293–299.

Ramanathan, V., B. R. Barkstrom, and E. F. Harrison, 1989a: Climate and the Earth's radiation budget. *Physics Today,* **42(5),** 22–32.

Ramanathan, V., R. D. Cess, E. F. Harrison, P. Minnis, B. R. Barkstrom, E. Ahmad, and D. Hartman, 1989b: Cloud-radiative forcing and climate:

Results from the earth radiation budget experiment. *Science,* **243,** 57–63.

Ramanathan, V., R. J. Cicerone, H. B. Singh, and J. T. Kiehl, 1985: Trace gas trends and their potential role in climatic change. *Journal of Geophysical Research,* **90,** 5547–5566.

Ramanathan, V., and W. Collins, 1991: Thermodynamic regulation of ocean warming by cirrus clouds deduced from observations of the 1987 El Niño. *Nature,* **351,** 27–32.

Ramanathan, V., M. S. Lian, and R. D. Cess, 1979: Increased atmospheric CO_2: Zonal and seasonal estimates on the effect on the radiation energy balance and surface temperature. *Journal of Geophysical Research,* **84,** 4949–4958.

Ramanathan, V., E. J. Pitcher, R. C. Malone, and M. L. Blackmon, 1983: The response of a spectral GCM to refinements in radiative processes. *Journal of the Atmospheric Sciences,* **40,** 605–630.

Rampino, M. R., and S. Self, 1984: Sulphur-rich volcanic eruptions and stratospheric aerosols. *Nature,* **310,** 677–679.

Rasmussen, R. A., and M. A. K. Khalil, 1986: Atmospheric trace gases: Trends and distributions over the last decade. *Science,* **232,** 1623–1624.

Rasool, S. I., and S. H. Schneider, 1971: Atmospheric CO_2 and aerosols: Effects of large increases on global climate. *Science,* **173,** 138–141.

Raval, A., and V. Ramanathan, 1989: Observational determination of the greenhouse effect. *Nature,* **342,** 758–761.

Raynaud, D., and J. M. Barnola, 1985: An Antarctic ice core reveals atmospheric CO_2 variations over the past few centuries. *Nature,* **315,** 309–311.

Reid, G. C., 1991: Solar irradiance variations and the global sea surface temperature record. *Journal of Geophysical Research,* **96,** 2835–2844.

Riebsame, W. E., 1990: Anthropogenic climate change and a new paradigm of natural resource planning. *The Professional Geographer,* **42,** 1–12.

Rind, D., and M. Chandler, 1991: Increased ocean heat transports and warmer climate. *Journal of Geophysical Research,* **96,** 7437–7461.

Rind, D., E.-W. Chiou, W. Chu, J. Larsen, S. Oltmans, J. Lerner, M. P. McCormick, and L. McMaster, 1991: Positive water vapour feedback in climate models confirmed by satellite data. *Nature,* **349,** 500–503.

Rind, D., R. Goldberg, J. Hansen, C. Rosenzweig, and R. Ruedy, 1990: Potential evapotranspiration and the likelihood of future drought. *Journal of Geophysical Research,* **95,** 9983–10,004.

Rind, D., R. Goldberg, and R. Ruedy, 1989: Change in climate variability in the 21st century. *Climatic Change,* **14,** 5–37.

Roberts, L., 1991: Greenhouse role in reef stress unproven. *Science,* **253,** 258–259.

Roberts, W. O., and H. Lansford, 1979: *The Climate Mandate.* San Francisco: W. H. Freeman.

Robinson, D. A., and K. F. Dewey, 1990: Recent secular variations in the extent of Northern Hemispheric snow cover. *Geophysical Research Letters,* **17,** 1557–1560.

Rodhe, H., 1990: A comparison of the contribution of various gases to the greenhouse effect. *Science,* **248,** 1217–1219.

Roeckner, E., U. Schlese, J. Biercamp, and P. Loewe, 1987: Cloud optical depth and climate modelling. *Nature,* **329,** 138–140.

Romme, W. H., and D. G. Despain, 1989: The Yellowstone fires. *Scientific American,* **261,** 37–46.

Rotty, R. M., and C. D. Masters, 1985: Carbon dioxide from fossil fuel combustion: Trends, resources, and technological implications, *in* Trabalka, J.R. (editor), *Atmospheric Carbon Dioxide and the Global Carbon Cycle.* Washington, D.C.: U.S. Department of Energy, 63–80.

Royal Meteorological Society, 1991: The greenhouse effect—draft of a statement from the Royal Meteorological Society. *Weather,* **46,** 26–29.

Royer, J. F., S. Planton, and M. Deque, 1990: A sensitivity experiment for the removal of Arctic sea ice with the French spectral general circulation models. *Climate Dynamics,* **5,** 1–17.

Ruddiman, W. F., and J. E. Kutzbach, 1991: Plateau uplift and climatic change. *Scientific American,* **264,** 66–75.

Saffir, H. S., 1977: Design and construction requirements for hurricane resistance construction. New York: American Society of Civil Engineers, Preprint 2830.

Sansom, J., 1989: Antarctic surface temperature time series. *Journal of Climate,* **2,** 1164–1172.

Schindler, D. W., K. G. Beaty, E. J. Fee, D. R. Cruikshank, E. R. DeBruyn, D. L. Findlay, G. A. Linsey, J. A. Shearer, M. P. Stainton, and M. A. Turner, 1990: Effects of climatic warming on lakes of the central boreal forest. *Science,* **250,** 967–970.

Schlesinger, M. E., and X. Jiang, 1988: The transport of CO_2-induced warming into the ocean: An analysis of simulations by the OSU coupled atmosphere-ocean general circulation model. *Climate Dynamics,* **3,** 1–17.

———1990: Simple model representation of atmosphere-ocean GCMs and estimation of the time scale of CO_2-induced climate change. *Journal of Climate,* **3,** 1297–1315.

———1991: Revised projection of future greenhouse warming. *Nature,* **350,** 219–221.

Schlesinger, M. E., and J. F. B. Mitchell, 1987: Climatic model simulations of the equilibrium climate response to increased carbon dioxide. *Reviews in Geophysics,* **25,** 760–798.

Schlesinger, W. H., J. F. Reynolds, G. L. Cunningham, L. F. Huenneke, W. M. Jarrell, R. A. Virginia, and W. G. Whitford, 1990: Biological feedbacks in global desertification. *Science,* **247,** 1043–1048.

Schmitt, C., and D. A. Randall, 1991: Effects of surface temperature and clouds on the CO_2 forcing. *Journal of Geophysical Research,* **96,** 9159–9168.

Schneider, S. H., 1975: On the carbon dioxide–climate confusion. *Journal of the Atmospheric Sciences,* **32,** 2060–2066.

——1976: *The Genesis Strategy.* New York: Plenun Press.

——1989a: *Global Warming: Are We Entering the Greenhouse Century?* San Francisco: Sierra Club Books.

——1989b: The greenhouse effect: Science and policy. *Science,* **243,** 771–781.

——1990: The global warming debate heats up: An analysis and perspective. *Bulletin of the American Meteorological Society,* **71,** 1292–1304.

Schneider, S. H., and R. E. Dickinson, 1974: Climate modeling. *Reviews of Geophysics and Space Physics,* **12,** 447–493.

Schneider, S. H., and S. L. Thompson, 1981: Atmospheric CO_2 and climate: Importance of the transient response. *Journal of Geophysical Research,* **86,** 3135–3147.

Schönwiese, C.-D., 1987: Moving spectral variance and coherence analysis and some applications on long air temperature series. *Journal of Climate and Applied Meteorology,* **26,** 1723–1730.

Schönwiese, C.-D., and J. Malcher, 1987: The CO_2 temperature response: A comparison of the results from general circulation models with statistical assessments. *Journal of Climatology,* **7,** 215–229.

Schönwiese, C.-D., and K. Runge, 1991: Some updated statistical assessments of the surface temperature response to increased greenhouse gases. *International Journal of Climatology,* **11,** 237–250.

Schwartz, S. E., 1988: Are global cloud cover and climate controlled by marine phytoplankton? *Nature,* **336,** 441–445.

Sear, C. B., P. M. Kelly, P. D. Jones, and C. M. Goodess, 1987: Global surface-temperature responses to major volcanic eruptions. *Nature,* **330,** 365–367.

Seaver, W. L., and J. E. Lee, 1987: A statistical examination of sky cover changes in the contiguous United States. *Journal of Climate and Applied Meteorology,* **26,** 88–95.

Sedlacek, W. A., E. J. Mroz, A. L. Lazrus, and B. W. Gandrud, 1983: A decade of stratospheric sulphate measurements compared with observations of volcanic eruptions. *Journal of Geophysical Research,* **88,** 3741–3776.

Seitz, F., R. Jastrow, and W. A. Nierenberg, 1989: *Scientific Perspectives on the Greenhouse Problem.* Washington, D.C.: George C. Marshall Institute.

Self, S., M. R. Rampino, and J. J. Barbera, 1981: The possible effects of large 19th and 20th century volcanic eruptions on zonal and hemispheric surface temperatures. *Journal of Volcanology and Geothermal Research,* **11,** 41–60.

Sellers, W. D., 1965: *Physical Climatology.* Chicago: University of Chicago Press.

———1969: A global climatic model based on the energy balance of the earth-atmosphere system. *Journal of Applied Meteorology,* **8,** 392–400.

———1973: A new global climatic model. *Journal of Applied Meteorology,* **12,** 241–254.

Sellers, W. D., and W. Liu, 1988: Temperature patterns and trends in the upper troposphere and lower stratosphere. *Journal of Climate,* **1,** 573–581.

Shackleton, N. J., and N. D. Opdyke, 1973: Oxygen isotope and paleo-magnetic stratigraphy of equatorial Pacific core V28-238: Oxygen isotope temperatures and ice volumes on a 10^5 and 10^6 year scale. *Quaternary Research,* **3,** 39–55.

Shine, K. P., R. G. Derwent, D. J. Wuebbles, and J.-J. Morcrette, 1990: Radiative forcing of climate, *in* Houghton, J. T., G. J. Jenkins, and J. J. Ephraums (editors), *Climate Change: The IPCC Scientific Assessment.* Cambridge, England: Cambridge University Press, 41–68.

Shukla, J., C. Nobre, and P. Sellers, 1990: Amazon deforestation and climate change. *Science,* **247,** 1322–1325.

Singer, S. F. (editor), 1989: *Global Climate Change.* New York: Paragon House.

Slingo, A., 1989: Wetter clouds dampen global greenhouse warming. *Nature,* **341,** 104.

———1990: Sensitivity of the earth's radiation budget to changes in low clouds. *Nature,* **343,** 49–51.

Smagorinsky, J., 1974: Global atmospheric modeling and the numerical simulation of climate, *in* Hess, W. N. (editor), *Weather and Climate Modification.* New York: John Wiley & Sons, 633–686.

Smith, J. B., 1991: The potential impacts of climate change on the Great Lakes. *Bulletin of the American Meteorological Society,* **72,** 21–28.

Smith, K. A., 1990: Greenhouse gas fluxes between land surfaces and the atmosphere. *Progress in Physical Geography,* **14,** 349–372.

Smith, L. D., and T. H. Vonder Haar, 1991: Cloud-radiation interactions in a general circulation model: Impact upon the planetary radiation balance. *Journal of Geophysical Research,* **96,** 893–914.

Solow, A. R., and J. M. Broadus, 1989: On the detection of greenhouse warming. *Climatic Change,* **15,** 449–453.

Somerville, R. C., and L. A. Remer, 1984: Cloud optical thickness feed-backs in the CO_2 climate problem. *Journal of Geophysical Research,* **89,** 9668–9672.

Spelman, M. J., and S. Manabe, 1984: Influence of oceanic heat transport upon the sensitivity of a model climate. *Journal of Geophysical Research,* **89,** 571–586.

Spencer, R. W., and J. R. Christy, 1990: Precise monitoring of global temperature trends from satellites. *Science,* **247,** 1558–1562.

Spencer, R. W., J. R. Christy, and N. C. Grody, 1990: Global atmospheric temperature monitoring with satellite microwave measurements: Method and results 1979–84. *Journal of Climate,* **3,** 1111–1128.

Staffelbach, T., B. Stauffer, A. Sigg, and H. Oeschger, 1991: CO_2 measurements from polar ice cores: More data from different sites. *Tellus,* **43B,** 91–96.

Stanton, B. R., 1991: Ocean circulations and ocean-atmosphere exchanges. *Climatic Change,* **18,** 175–194.

Staubes, R., H.-W. Georgii, and G. Ockelmann, 1989: Flux of CO_2, DMS and CS_2 from various soils in Germany. *Tellus,* **41B,** 305–313.

Stauffer, B., G. Fischer, A. Neftel, and H. Oeschger, 1985: Increase of atmospheric methane recorded in Antarctic ice core. *Science,* **229,** 1386–1388.

Stephens, G. L., and T. J. Greenwald, 1991: The earth's radiation budget and its relation to atmospheric hydrology. 1. Observations of the clear sky greenhouse effect. *Journal of Geophysical Research,* **96,** 15,311–15,324.

Stone, P. H., and J. S. Risbey, 1990: On the limitations of general circulation climate models. *Geophysical Research Letters,* **17,** 2173–2176.

Stouffer, R. J., S. Manabe, and K. Bryan, 1989: Interhemispheric asymmetry in climate response to a gradual increase of atmospheric CO_2. *Nature,* **342,** 660–662.

Sun, M., 1989: Global warming becomes hot issue for Bromley. *Science,* **246,** 569.

Synthesis Panel, Committee on Science, Engineering, and Public Policy, National Academy of Sciences, 1991: *Policy Implications of Greenhouse Warming.* Washington, D.C.: National Academy Press.

Sze, N. D., 1977: Anthropogenic CO emissions: Implications for the atmospheric CO-OH-CH_4 cycle. *Science,* **195,** 673–675.

Tangley, L., 1988: Preparing for climate change. *BioScience,* **38,** 14–18.

Tans, P. P., T. J. Conway, and T. Nakasawa, 1989: Latitudinal distribution of sources and sinks of atmospheric carbon dioxide. *Journal of Geophysical Research,* **94,** 5151–5172.

Tans, P. P., I. Y. Fung, and T. Takahashi, 1990: Observational constraints on the global atmospheric CO_2 budget. *Science,* **247,** 1431–1438.

Thiemens, M. H., and W. C. Trogler, 1991: Nylon production: An unknown source of atmospheric nitrous oxide. *Science,* **251,** 932–934.

Thompson, R. D., 1989: Short-term climatic change: Evidence, causes, environmental consequences and strategies for action. *Progress in Physical Geography,* **13,** 315–347.

Thoning, K. W., P. P. Tans, and W. D. Komhyr, 1989: Atmospheric carbon dioxide at Mauna Loa Observatory 2. Analysis of the NOAA GMCC data, 1974–1985. *Journal of Geophysical Research,* **94,** 8549–8565.

Tickell, C., 1977: *Climatic Change and World Affairs.* Cambridge, Massachusetts: Harvard University Center for International Affairs.

Trabalka, J. R. (editor), 1985: *Atmospheric Carbon Dioxide and the Global Carbon Cycle.* Washington, D.C.: U.S. Department of Energy, DOE/ER-0239.

Trenberth, K. E., 1990: Recent observed interdecadal climate changes in the Northern Hemisphere. *Bulletin of the American Meteorological Society,* **71,** 988–993.

Trenberth, K. E., and J. G. Olson, 1989: Temperature trends at the South Pole and McMurdo Sound. *Journal of Climate,* **2,** 1196–1206.

Tricot, C., and A. Berger, 1987: Modelling the equilibrium and transient response of global temperature to past and future trace gas concentrations. *Climate Dynamics,* **2,** 39–61.

Tsonis, A. A., and J. B. Elsner, 1989: Testing the global warming hypothesis. *Geophysical Research Letters,* **16,** 795–797.

Tyndall, J., 1861: On the absorption and radiation of heat by gases and vapours, and on the physical connexion of radiation, absorption, and conduction. *Philosophical Magazine,* **22,** 169–194, 273–285.

Van der Veen, C. J., 1988: Projecting future sea level. *Surveys in Geophysics,* **9,** 389–418.

Victor, D. G., 1991: How to stop global warming. *Nature,* **349,** 451–456.

Vinnikov, K. Y., P. Y. Groisman, and K. M. Lugina, 1990: Empirical data on contemporary global climate changes (temperature and precipitation). *Journal of Climate,* **3,** 662–667.

Wadhams, P., 1990: Evidence for thinning of the Arctic ice cover north of Greenland. *Nature,* **345,** 795–797.

Wahlen, M., D. Allen, B. Deck, and A. Herchenroder, 1991: Initial measurements of CO_2 concentrations (1530 to 1940 AD) in air occluded in the GISP 2 ice core from central Greenland. *Geophysical Research Letters,* **18,** 1457–1460.

Walsh, J. E., 1991: The Arctic as a bellwether. *Nature,* **352,** 19–20.

Wang, W.-C., M. P. Dudek, X.-Z. Liang, and T. J. Kiehl, 1991: Inadequacy of effective CO_2 as a proxy in simulating the greenhouse effect of other radiatively active trace gases. *Nature,* **350,** 573–577.

Wang, W.-C., Y. L. Yung, A. A. Lacis, T. Mo, and J. E. Hansen, 1976: Greenhouse effects due to man-made perturbations of trace gases. *Science,* **194,** 685–690.

Wang, W.-C., Z. Zhaomei, and T. R. Karl, 1990: Urban heat islands in China. *Geophysical Research Letters,* **17,** 2377–2389.

Warren, S. G., C. J. Hahn, J. London, R. M. Chervin, and R. Jenne, 1988: *Global Distribution of Total Cloud Cover and Cloud Type Amounts over the Ocean.* Boulder, Colorado: U.S. Department of Energy and National Center for Atmospheric Research.

Warrick, R., and J. Oerlemans, 1990: Sea level rise, *in* Houghton, J. T., G. J. Jenkins, and J. J. Ephraums (editors), *Climate Change: The IPCC Scientific Assessment.* Cambridge, England: Cambridge University Press, 257–281.

Warrick, R. A., and T. M. L. Wigley (editors), 1990: *Climate and Sea Level Change.* Cambridge, England: Cambridge University Press.

Washington, W. M., and G. A. Meehl, 1983: General circulation model experiments on the climatic effects due to a doubling and quadrupling of carbon dioxide concentration. *Journal of Geophysical Research,* **88,** 6600–6610.

———1984: Seasonal cycle experiments on the climate sensitivity due to doubling of CO_2 with an atmospheric general circulation model coupled with a mixed-layer ocean model. *Journal of Geophysical Research,* **89,** 9475–9503.

———1989: Climate sensitivity due to increased CO_2: Experiments with a coupled atmosphere and ocean general circulation model. *Climate Dynamics,* **4,** 1–38.

Watson, A. J., C. Robinson, J. E. Robinson, P. J. le B. Williams, and M. J. R. Fasham, 1991: Spatial variability in the sink for atmospheric carbon dioxide in the North Atlantic. *Nature,* **350,** 50–53.

Watson, R. T., H. Rodhe, H. Oeschger, and U. Siegenthaler, 1990: Greenhouse gases and aerosols, *in* Houghton, J. T., G. J. Jenkins, and J. J. Ephraums (editors), *Climate Change: The IPCC Scientific Assessment.* Cambridge, England: Cambridge University Press, 1–40.

Watts, R. G., 1980: Climate models and CO_2-induced climatic changes. *Climatic Change,* **2,** 387–408.

Watts, R. G., and M. C. Morantine, 1991: Is the greenhouse gas-climate signal hiding in the deep ocean? *Climatic Change,* **18,** iii–vi.

Weber, G.-R., 1990: Tropospheric temperature anomalies in the Northern Hemisphere 1977–1986. *International Journal of Climatology,* **10,** 3–19.

———1991: Long-term decline in the frequency of gale-force winds in West Germany. *Theoretical and Applied Climatology,* **44,** 43–46.

Wetherald, R. T., and S. Manabe, 1981: Influence of seasonal variation upon the sensitivity of a model climate. *Journal of Geophysical Research,* **86,** 1194–1204.

———1986: An investigation of cloud cover change in response to thermal forcing. *Climatic Change,* **8,** 5–23.

————1988: Cloud feedback processes in a GCM. *Journal of the Atmospheric Sciences,* **45,** 1397–1415.

White, R. M., 1989: Greenhouse policy and climate uncertainty. *Bulletin of the American Meteorological Society,* **70,** 1123–1127.

Wigley, T. M. L., 1987: Relative contributions of different trace gases to the greenhouse effect. *Climate Monitor,* **16,** 14–28.

————1989: Possible climate change due to SO_2-derived cloud condensation nuclei. *Nature,* **338,** 365–367.

————1991: Could reducing fossil-fuel emissions cause global warming? *Nature,* **349,** 503–506.

Wigley, T. M. L., and T. P. Barnett, 1990: Detection of the greenhouse effect in the observations, *in* Houghton, J. T., G. J. Jenkins, and J. J. Ephraums (editors), *Climate Change: The IPCC Scientific Assessment.* Cambridge, England: Cambridge University Press, 239–255.

Wigley, T. M. L., and P. D. Jones, 1981: Detecting CO_2-induced climate change. *Nature,* **292,** 205–208.

————1988: Do large-area-average temperature series have an urban warming bias? *Climatic Change,* **12,** 313–319.

Wigley, T. M. L., P. D. Jones, and P. M. Kelly, 1980: Scenario for a warm, high-CO_2 world. *Nature,* **283,** 17–21.

Wigley, T. M. L., and S. C. B. Raper, 1990a: Climatic change due to solar irradiance changes. *Geophysical Research Letters,* **17,** 2169–2172.

————1990b: Natural variability of the climate system and detection of the greenhouse effect. *Nature,* **344,** 324–327.

Wigley, T. M. L., and M. E. Schlesinger, 1985: Analytical solution for the effect of increasing CO_2 on global mean temperature. *Nature,* **315,** 649–652.

Williams, J., 1980: Anomalies in temperature and rainfall during warm Arctic seasons as a guide to the formulations of climate scenarios. *Climatic Change,* **2,** 249–266.

Wilson, C. A., and J. F. B. Mitchell, 1987: A doubled CO_2 climate sensitivity experiment with a global climate model including a simple ocean. *Journal of Geophysical Research,* **92,** 13,315–13,342.

Wood, F. B., 1988a: Comment: On the need for validation of the Jones et al. temperature trends with respect to urban warming. *Climatic Change,* **12,** 297–312.

————1988b: Global alpine glacier trends, 1960s to 1980s. *Arctic and Alpine Research,* **20,** 404–413.

————1990: Monitoring global climate change: The case of greenhouse warming. *Bulletin of the American Meteorological Society,* **71,** 42–52.

Woodwell, G. M., 1989: The warming of the industrialized middle latitudes 1985–2050: Causes and consequences. *Climatic Change,* **15,** 31–50.

Wright, P. B., 1985: The Southern Oscillation: An ocean-atmosphere feedback system. *Bulletin of the American Meteorological Society,* **66,** 398–412.

Wu, Z., R. E. Newell, and J. Hsiung, 1990: Possible factors controlling global marine temperature variations over the past century. *Journal of Geophysical Research,* **95,** 11,799–11,810.

INDEX

ABOUT THE AUTHOR

Robert C. Balling, Jr. is director of the Office of Climatology and associate professor of geography at Arizona State University.

PACIFIC RESEARCH INSTITUTE FOR PUBLIC POLICY

OTHER STUDIES IN PUBLIC POLICY BY
THE PACIFIC RESEARCH INSTITUTE

URBAN TRANSIT
The Private Challenge to Public Transportation
Edited by Charles A. Lave
Foreword by John Meyer

POLITICS, PRICES, AND PETROLEUM
The Political Economy of Energy
By David Glasner
Foreword by Paul W. MacAvoy

RIGHTS AND REGULATION
Ethical, Political, and Economic Issues
Edited by Tibor M. Machan and M. Bruce Johnson
Foreword by Aaron Wildavsky

FUGITIVE INDUSTRY
The Economics and Politics of Deindustrialization
By Richard B. McKenzie
Foreword by Finis Welch

MONEY IN CRISIS
The Federal Reserve, the Economy, and Monetary Reform
Edited by Barry N. Siegel
Foreword by Leland B. Yeager

NATURAL RESOURCES
Bureaucratic Myths and Environmental Management
By Richard Stroup and John Baden
Foreword by William Niskanen

FIREARMS AND VIOLENCE
Issues of Public Policy
Edited by Don B. Kates, Jr.
Foreword by John Kaplan

WATER RIGHTS
Scarce Resource Allocation, Bureaucracy, and the Environment
Edited by Terry L. Anderson
Foreword by Jack Hirshleifer

LOCKING UP THE RANGE
Federal Land Controls and Grazing
By Gary D. Libecap
Foreword by Jonathan R. T. Hughes

THE PUBLIC SCHOOL MONOPOLY
A Critical Analysis of Education and the State in American Society
Edited by Robert B. Everhart
Foreword by Clarence J. Karier

RESOLVING THE HOUSING CRISIS
Government Policy, Demand, Decontrol, and the Public Interest
Edited with an Introduction by M. Bruce Johnson

OFFSHORE LANDS
Oil and Gas Leasing and Conservation on the Outer Continental Shelf
By Walter J. Mead, et al.
Foreword by Stephen L. McDonald